SCIENCE AND SPIRITUALITY

Making Room for Faith in the Age of Science

In *Science and Spirituality: Making Room for Faith in the Age of Science*, Michael Ruse offers a new analysis of the often troubled relationship between science and religion. Arguing against both extremes – in one corner, the New Atheists; in the other, the Creationists and their offspring, the Intelligent Designers – he asserts that science is undoubtedly the highest and most fruitful source of human inquiry. Yet, by its very nature and its deep reliance on metaphor, science restricts itself and is unable to answer basic, significant, and potent questions about the meaning of the universe and humankind's place within it: Why is there something rather than nothing? What is the ultimate source and foundation of morality? What is the nature of consciousness? What is the meaning of it all? Ruse shows that one can legitimately be a skeptic about all of these questions, yet why it is nonetheless open to a Christian, or a member of any faith, to offer answers. Scientists, he concludes, should be proud of their achievements but modest about their scope. Christians should be confident of their mission but respectful of the successes of science.

Michael Ruse is the Lucyle T. Werkmeister Professor of Philosophy and Director of the Program in the History and Philosophy of Science at Florida State University. The author or editor of numerous books, most recently *Darwinism and Its Discontents* and *The Cambridge Companion to the "Origin of Species"* (with Robert Richards), he has been a Herbert Spencer Lecturer at Oxford University, a Gifford Lecturer at the University of Glasgow, and Reynolds Lecturer at Baylor University. He is a Fellow of the Royal Society of Canada and of the American Association for the Advancement of Science and the recipient of three honorary degrees.

To Ronald Numbers

SCIENCE AND SPIRITUALITY

Making Room for Faith in the Age of Science

Michael Ruse

Florida State University

CAMBRIDGE UNIVERSITY PRESS
Cambridge, New York, Melbourne, Madrid, Cape Town, Singapore,
São Paulo, Delhi, Dubai, Tokyo

Cambridge University Press
32 Avenue of the Americas, New York, NY 10013-2473, USA

www.cambridge.org
Information on this title: www.cambridge.org/9780521755948

First published 2010

Printed in the United States of America

A catalog record for this publication is available from the British Library.

Library of Congress Cataloging in Publication data
Ruse, Michael.
Science and spirituality : making room for faith in the age of science / Michael Ruse.
p.   cm.
Includes bibliographical references and index.
ISBN 978-0-521-75594-8 (hardback)
1. Religion and science.   I. Title
BL240.3.R875   2010
261.5'5–dc22   2009020206

ISBN 978-0-521-75594-8 Hardback

CONTENTS

v

ACKNOWLEDGMENTS

In writing this book, I have learned from and been encouraged by many good friends and colleagues, most of whom will reject the conclusion that I draw. These include Francisco J. Ayala, John Beatty, Joe Cain, David Castle, Fritz Davis, Wim Drees, Peter Harrison, John Haught, Philip Hefner, David Livingstone, Ronald Numbers, Robert J. Richards, Joseph Travis, and Edward O. Wilson. Also I must acknowledge two great influences, now deceased: Jay Newman and Arthur Peacocke. I am particularly in the debt of two members of the Florida State University Department of Religion, Matthew Day and John Kelsay. They have given generously of their time and knowledge, the former especially with respect to history and the latter to theology. Above all others, two very longtime friends hover above the project – Mary Hesse, who taught me the importance of metaphor in science, and Ernan McMullin, who showed me that one can, as a philosopher, write profitably and responsibly on the science-religion relationship. Samantha Muka and Peter Takacs spent many hours in the library searching down books and articles. As always, I am grateful to the encouraging and professional folk at Cambridge University Press, my editor, Beatrice Rehl, and my copy editor, Russell Hahn. Kathleen Paparchontis did the index. And last but certainly not least, I record my thanks to and love for my wife, Lizzie.

Support for this project came from the very generous gift to the Philosophy Department of Florida State University by William and Lucyle Werkmeister. Given the social context in which this book is appearing, it is perhaps worth saying that this was the only source of support and that there are absolutely no conditions attached to the gift. The agenda and conclusions are mine and mine alone.

INTRODUCTION

Steven Weinberg, winner of the Nobel Prize in physics for combining electromagnetism and the weak force into the electroweak force, has a deservedly high profile. He is respected not just as a very good scientist but also as one who is able to communicate the most technical aspects of his work to the general public. His book *The First Three Minutes* (1977), about the early history of our universe, is a dazzling display of expertise combined with a brilliant ability to explain complex ideas without at any point trivializing them or condescending to his audience. Rightfully, he has assumed the mantle of a leading spokesperson for science. When he speaks, he speaks with authority. One should therefore take him seriously when he speaks, as a scientist, about so important a topic as religion. And spoken he certainly has. In 1999, in a dialog on religion with another scientist (Sir John Polkinghorne), he said: "Religion is an insult to human dignity. Without it you would have good people doing good things and evil people doing evil things. But for good people to do evil things, that takes religion." Also speaking as a scientist, he tells us: "This is one of the great social functions of science – to free people from superstition" (see Weinberg 2001, 231, for a printed version). Let there be no mistake, the greatest superstition of them all is religion. It is here that science has its most powerful cleansing action: "It's a consequence of the experience of science. As you learn more and more about the universe, you find you can understand more and more without any reference to supernatural intervention, so you lose interest in that possibility. Most scientists I know don't care enough about religion even to call themselves atheists. And that, I think, is one of the great things about science – that it has made it possible for people not to be religious" (Angier 2001, 37). Not only does science refute religion, it makes it basically boring.

There are Nobel laureates even more famous than Weinberg who share his sentiments entirely. Thanks to his discovery of the double helix, Francis Crick was one of the iconic figures of the twentieth century. He has written: "If revealed religions have revealed anything it is that they are usually wrong" (Crick 1994, 258). And elsewhere: "A knowledge of the true age of the earth and of the fossil record makes it impossible for any balanced intellect to believe in the literal truth of every part of the Bible in the way that fundamentalists do. And if some of the Bible is manifestly wrong, why should any of the rest of it be accepted automatically?" He added that "it is clear that some mysteries have still to be explained scientifically. While these remain unexplained, they can serve as an easy refuge for religious superstition. It seemed to me of the first importance to identify these unexplained areas of knowledge and to work toward their scientific understanding whether such explanations would turn out to confirm existing beliefs or to refute them" (Crick 1988, 11). Revealingly, this quote came from that section of Crick's autobiography where he tells us that he is "an agnostic with a strong inclination toward atheism" (p. 10). Crick's codiscoverer of the double helix, James Watson, holds very similar sentiments: "I don't think we're for anything, we're just products of evolution. You can say 'Gee, your life must be pretty bleak if you don't think there's a purpose' but I'm anticipating a good lunch." On being asked if he knew any religious scientists: "Virtually none. Occasionally I meet them and I'm a bit embarrassed because I can't believe that anyone accepts truth by revelation" (BBC film interview with Richard Dawkins, 1996).

These are the heavyweights of science. The popularizers have no such claims to great achievement, but in their way they are even more important in forming the public's opinion about science and its relationship to the rest of our culture, including – especially including – religion. Here the hostility of science toward religion – toward the Christian religion – moves from the strong to the frenetic. Today's most popular science writer is the English biologist Richard Dawkins. He has written one smash-hit best-seller after another. Through them all, not to mention in numerous occasional pieces and interviews, one sees a growing hostility to religion – a hostility that Dawkins roots in his love and respect for science: "Faith is the great cop-out, the great excuse to evade the need to think and evaluate evidence. Faith is belief in spite of, even perhaps because of, the lack of evidence" (Richard Dawkins, untitled lecture, Edinburgh Science Festival, 1992). "I am against religion because it teaches us to be satisfied with not understanding the world" (RichardDawkins.net Quote 49). "It is fashionable to wax apocalyptic about the threat to humanity posed by the AIDS virus, 'mad cow' disease, and many others, but I think a case can be made that faith is one of the world's great evils, comparable to the smallpox virus but harder to eradicate" (Dawkins 1997, 2b). We reach the apotheosis in his most recent book, *The God Delusion*: "Once, in the question time after a lecture in Dublin, I was asked what I thought about the widely publicized cases of sexual abuse by Catholic priests in Ireland. I replied that, horrible as sexual abuse no doubt was, the damage was arguably less than the long-term psychological damage inflicted by bringing the child up Catholic in the first place" (Dawkins 2007, 316).

You might protest, with reason, that scientists are not always the best people to discern and understand the full implications of their own work. We must defer to them as scientists, but as thinkers about the broader implications of what they produce we should turn to others. We should turn to those whose professional expertise is directed toward the nature of science and how it relates to the nonscientific, including religion. We should turn to the philosophers and historians of science. Prepare for disappointment if you expect to hear another song. The philosopher Daniel Dennett, a former president of the American Philosophical Association, well known for his provocative and sparkling works on cognitive science and more recently on Darwinian evolutionary theory, makes no bones about his dislike of religion and his belief that science can and should sweep away such pernicious nonsense. "If religion isn't the greatest threat to rationality and scientific progress, what is? Perhaps alcohol, or television, or addictive video games. But although each of these scourges – mixed blessings, in fact – has the power to overwhelm our best judgment and cloud our critical faculties, religion has a feature of that none of them can boast: it doesn't just disable, it honours the disability." He sneers that: "People are revered for their capacity to live in a dream world, to shield their minds from factual knowledge and make the major decisions of their lives by consulting voices in their heads that they call forth by rituals designed to intoxicate them." Concluding: "Our motto should be: Friends don't let friends steer their lives by religion" (Dennett and Winston 2008, 14).

Fellow philosopher Philip Kitcher dwells on the suffering brought on by the struggle for existence, the prelude to Charles Darwin's mechanism of evolutionary change, natural selection. He writes:

> [George John] Romanes and [William] James, like the evangelical Christians who rally behind intelligent design today, appreciate that Darwinism is subversive. They recognize that the Darwinian picture of life is at odds with a particular kind of religion, Providentialist religion, as I shall call it. A large number of Christians, not merely those who maintain that virtually all of the Bible must be read literally, are providentialists. For they believe that the universe has been created by a Being who has a great design, a Being who cares for his creatures, who observes the fall of every sparrow and who is especially concerned with humanity. Yet the story of a wise and loving Creator, who has planned life on earth, letting it unfold over four billion years by the processes envisaged in evolutionary theory, is hard to sustain when you think about the details. (Kitcher 2007, 122–3)

He writes of having believed that Darwinism was reconcilable with Christianity and – with the fervor of a repenting sinner at an evangelical revivalist meeting – insists that he alone should be held responsible for "the earlier errors that I recant here" (p. 180).

The distinguished historian of science William Provine has spent a lifetime trying to atone for his Christian childhood.

> Of course, it is still possible to believe in both modern evolutionary biology and a purposive force, even the Judeo-Christian God. One can suppose that God started the whole universe or works through the laws of nature (or both). There is no contradiction between this or similar views of God and natural selection. But this view of God is also worthless. Called Deism in the seventeenth and eighteenth centuries and considered equivalent to atheism then, it is no different now. A God or purposive force that merely starts the universe or works through the laws of nature has nothing to do with

human morals, answers no prayers, gives no life everlasting, in fact does nothing whatsoever that is detectable. In other words, religion is compatible with modern evolutionary biology (and indeed all of modern science) if the religion is effectively indistinguishable from atheism. (Provine 1987, 51–2)

Provine thinks that you have to check your brains at the door of the church and that the only reason why scientists in America do not speak out is because they fear the loss of their grants, a real possibility in a society where so many of the citizens and their political leaders are practicing Christians. "Rather than simple intellectual dishonesty, this position is pragmatic" (p. 52).

It is true that not everyone subscribes to the "warfare" thesis embraced by these various people of science. Unfortunately, the proposed alternatives run the gamut from the barely honest, through the inept and gullible, to the banal and inadequate. Before his too-early death in 2002, the Harvard paleontologist Stephen Jay Gould was as much a public figure of science as Dawkins is now. His pronouncements on science and culture were taken by many as gospel (to use a metaphor). His fondness for baseball and his fascination with its statistics were considered proof that America's national pastime is truly God-given (to use another metaphor). There was even an episode of *The Simpsons* devoted to him (at which point the metaphors are exhausted). He was not just a public figure for science but was prepared to use his time and energies in its defense, appearing as an expert witness for the American Civil Liberties Union in Arkansas in 1981, when there was a successful attack on a law mandating the teaching of Biblical literalism in the biology classrooms of the publicly funded schools of the state. In one of his later books, *Rocks of Ages*, Gould supposedly defended the integrity of religion in the face of science, declaring them to be noncompeting world pictures or (to use his term) "Magisteria." However, when you started to read the fine print, you soon discovered that Gould missed his calling as an illusionist. You had better not make too many claims about God or the Trinity or the Resurrection or any of those other extraneous eruptions on the fair face of religion. Claims of this kind fall into the domain of science: "what the universe is made of (fact) and why does it work this way (theory)." Stick to morality, a kind of sentiment with vaguely mystical connotations, and you will do just fine. Talk of miracles and that sort of thing is just silly. When Arthur

Peacocke, biochemist, Anglican priest, and winner of the Templeton Prize for Advances in Religion, presumed to suggest that God creates continuously through evolution rather than in one fell swoop, his God was ridiculed as one "retooling himself in the spiffy language of modern science" (Gould 1999, 217). Truly, Gould was happier joining the chorus of the scientists I have just introduced. We humans "are here because one odd group of fishes had a peculiar fin anatomy that could transform into legs for terrestrial creatures; because the earth never froze entirely during an ice age; because a small and tenuous species, arising in Africa a quarter of a million years ago, has managed, so far, to survive by hook and by crook. We may yearn for a higher answer – but none exists" (Stephen Jay Gould, *Life* magazine, December 1988, quoted in Haught 1996, 290). On another occasion, writing of the way in which a comet hit the earth 65 million years ago, thus eliminating the dinosaurs and making possible the rise of the mammals, he quipped: "In an entirely literal sense, we owe our existence, as large and reasoning mammals, to our lucky stars" (Gould 1989, 318).

The Reverend Professor Sir John Polkinghorne, another Anglican priest, another Templeton Prize winner, physicist, Fellow of the Royal Society, and debating partner of Steven Weinberg, has labored long to show the compatibility of science and religion. He writes that although "there is a feeling abroad that somehow science and religion are opposed to each other," in fact "science and religion seem to me to have in common that they are both exploring aspects of reality. They are capable of mutual interaction which, though at times it is puzzling, it can also be fruitful" (Polkinghorne 1986, xi). Unfortunately, Polkinghorne's writing soon descends from disinterested analyses of the science-religion relationship into apologies for thinking that fits more comfortably into the culture of Elizabeth the First than Elizabeth the Second. He may not have faith enough to raise Lazarus from the dead, but he does a pretty good job on pulling out the stakes from the hearts of arguments long buried by such philosophers as David Hume and Immanuel Kant. He has abandoned the quest to defend religion in the face of science for dubious excursions into showing that science really has Meaning, with a capital M. Modern physics, particularly, reveals the face of the Creator – as in the case of the Shroud of Turin, it is all just a question of suspending judgment and looking hard enough. Those constants – the force of gravity, the speed of light, the attractions between molecules, and so forth – cannot be just chance. They point to something, and that something is upward.

Then, again, there is Philip Kitcher. He may think that science destroys traditional belief, but he still refers to himself as a "spiritual Christian."

> Spiritual Christians abandon almost all of the standard stories about the life of Jesus. They give up on the extraordinary birth, the miracles, the literal resurrection. What survive are the teachings, the precepts and parables, and the eventual journey to Jerusalem and the culminating moment of the Crucifixion. That moment of suffering and sacrifice is seen, not as the prelude to some triumphant return and the promise of eternal salvation – all that, to repeat, is literally false – but as a symbolic presentation of the importance of compassion and of love without limits. We are to recognize our own predicament, the human predicament, through the lens of the man on the cross. (Kitcher 2007, 152–3)

The trouble with this, of course, is that it may indeed do for Kitcher – and perhaps for Gould. It may do for humanists and even for Unitarians, not to mention an assortment of Episcopelian bishops. It will not do for Christians. They really are providentialists. They do not want to "abandon almost all of the standard stories about the life of Jesus." Many of them want a Virgin Birth, miracles, and a literal Resurrection. All want the "triumphant return and the promise of eternal salvation." If getting on with science means that you cannot have these things, then so much the worse for science.

It is people like this who have set the question I shall tackle in this book. I am fully aware that there are those on the side of religion who are no less hostile to science. What I have to say will be as pertinent to these people as to any, but in this book my focus is on the people of science – among whom I would include myself. I accept and love science as much as any of the men I have just introduced. I spent the early part of my life learning about science and have now devoted all of my adult life to exploring it, expounding it, defending it. However, I am far from thinking that science is all there is in life, and it is this belief that motivates me here. I want to ask about the nature of science and about its limits. I want to see if indeed science is truly so antithetical to religious thinking. I want to see if, rather, one can be both a scientist or lover of science and, with integrity, a person of religion – more particularly, because I can do only so many things at the same time, if you can cherish science and its achievements and be a Christian, holding with dignity and proper conviction the things that are central to that religion.

I take seriously the notion of "things that are central to that religion." For this reason, this is the first and last you will read here of biblical literalism. I am fully aware that many, probably the majority, of Americans believe in the Bible taken absolutely literally – six days of creation, six thousand years ago, universal flood sometime shortly thereafter. Obviously claims like these are in conflict with modern science, as are the more specific claims of particular religions. Modern anthropology clashes categorically with the Mormon claim that the native people of America are the lost tribes of Israel. My position is simply that none of this is part of traditional Christianity, the Christianity of Augustine and Aquinas, Luther and Calvin. I am not saying that none of these beliefs was ever held, even by the people just listed, but I am saying that traditional Christianity has always insisted that truth cannot be opposed to truth, and that if modern science shows literal claims to be false, then these claims must be understood symbolically or metaphorically. Augustine insisted on this – the ancient Jews were nomadic people who did not have the scientific sophistication of educated Romans – and so too did Calvin, who spoke of God "accommodating" his language to the common people. Modern-day literalism, Creationism so-called, is an idiosyncratic legacy of nineteenth-century, American, evangelical Protestantism. Politically, it needs to be taken seriously. In a work trying to understand the relationship between science and the core of Christianity, the religion of the West of the past two thousand years, it can be ignored. I should say that in arguing thus I am being neither cavalier nor lazy. In several works I have discussed literalism in very great detail, both philosophically and historically, explaining why I find it less than adequate as a belief system. It is now time to talk of other things. (I discuss American biblical literalism in Ruse 1988, 2001, 2005, and 2008b, among other places.)

Lest you fear that I have a hidden agenda – perhaps, like John Polkinghorne, secretly wanting to proselytize, or, possibly worried about the threat to science in the United States posed by the forces of fundamentalist religion, about to offer a case based on political or (what Provine calls) pragmatic factors – I must at once emphatically tell you otherwise. You can rest assured that, for better or for worse, my intent is not some unspoken attempt to turn you to religious commitment. I am fully convinced that, whatever the nature of science, there is nothing there forcing you to be a Christian. You can also rest assured that the very last thing I want to do is to make a case for political or pragmatic reasons. I am certainly not going to make any arguments because I think they will play well in Peoria – or, more precisely, in the churches of Peoria. I am a philosopher, and, like most of my ilk, I am a bit of a snob about these sorts of things. I truly believe that the quest for understanding is in itself a noble enterprise. Expectedly, therefore, what does motivate me here, despite the fact that much of this book is more history than anything else, are some very traditional philosophical concerns. I am simply interested in the nature of science, its scope and its limits – and, in the light of this, in what the Christian can then legitimately say and claim. If things are genuinely in opposition to science, then so much the worse for them. But I believe that we do science a disservice by placing it in false opposition to other things in life and culture, especially other things that many people hold dear.

My approach to philosophy is that of the naturalist. My interest in limits does not belie my belief that the highest form of knowledge is scientific knowledge. I want to make my philosophy as much like science as possible. Where the scientist takes the physical world (including the organic world) as his or her datum, I take science as my datum. I am also an evolutionist. I do not much care for simple analogies between the history of organisms and the history of science, but I do believe that the key to understanding the present is understanding the past. Therefore, much of this book will be historical. It will not be a history of the science-religion relationship as such. Rather, it will be a history of science itself – very selective, obviously, but I trust sufficiently comprehensive to make the main points without too much fear of distortion because of biased focus. Only when this is done will I be able to turn to my philosophical analysis and see how, in the light of my findings, we should regard the Christian religion. I should say that a major reason for my restriction of the discussion to (Western) Christianity stems from my historical approach.It is against this religion that science has grown and defined itself in the five centuries since the beginning of the Scientific Revolution. I suspect that my conclusions can be generalized to other religions, but that is a task for another person at another time.

The subtitle of this book starts with reference to a comment made by Immanuel Kant in the Preface to the second edition of his masterwork, *The Critique of Pure Reason.* "I have therefore found it necessary to deny knowledge in order to make room for faith." I am not a Kantian. In many respects, I feel more comfortable with the philosophy of David Hume. I am certainly not going to deny knowledge as such, but then neither did Kant. He wanted to show its limitations, particularly the limitations of a science-based knowledge. That is my aim, too. The subtitle continues by adapting the name of a group of people (including me) interested in the science-religion relationship: The Institute on Religion in an Age of Science. Mainly liberal Christians, these people are as modest as the title of their group is pretentious. In labeling my book as I do, I show my respect and affection for men and women I am proud to call my friends. My dedication is to a man who has contributed more than any other to an understanding of science, of religion, and of their relationship. His writings and his friendship have made my life much richer than it would be otherwise.

ONE

THE WORLD AS AN
ORGANISM

Let us start at the beginning of Western science, something due to the Mesopotamians (living where we now find Iraq) and the Egyptians (Lindberg 1992). They both had various creation myths, usually involving gods and the moving of waters and land, with the sun and the moon having special significance. Not much distinction was drawn between what we would think of as scientific, causal understanding and what we would think of as magic. Often, contemporary social arrangements – like the status of priests and kings – were mixed up with all of this, and were traced back to happenings at the time of original creation. One thing of major importance was the development of writing, beginning with Egyptian hieroglyphics (picture symbols) about 3000 BCE and culminating in the Greek written language beginning about 800 BCE. Mathematics was important in both cultures. Starting around 3000 BCE, the Egyptians developed a decimal-based system, and from there went on to develop a very practically oriented mathematics – good for surveying and building and the like. It was not crude – they had rules, for instance, to work out the value of π – but it did not in any way compare to the sophistication of the Babylonian mathematics, which was, as is well known, sexagesimal (based on 60, thus giving the legacy of

11

our ways of measuring time and angle) as well as decimal. For the Babylonians, mathematics was less a tool for technology and more something vital to measuring the movements of the heavenly orbits, less for astronomy as such and more for making inferences about human destinies – that is, for astrology. The Babylonians were not cosmologists, that is, they weren't interested in the causes of the heavenly motions. For that, we turn to the Greeks.

GREEK SCIENCE IN ANTIQUITY

This is a vast topic, and I am selective, with an eye to my ends. We go back to around the sixth century BCE, and at once we start to notice a significant difference from the earlier ideas, namely, that although to the modern mind the speculations were often odd to the point of being ludicrous, the Greeks strove to give naturalistic answers, in the sense that they wanted things to be working according to normal laws and causes and – for all that we shall see significant roles for deities in some aspects – tried as much as possible to exclude the arbitrary whims or decisions of the gods in the workings of the world. An eclipse, for instance, had to have a natural explanation rather than simply being ascribed to the disfavor of some immortal being. In other words, we are dealing with a world of order, a cosmos, as opposed to a world of disorder, a chaos. Somewhat ahistorically, let us pick out three topics for discussion: mathematics, chemistry, and astronomy. As not directly relevant to our purpose, I leave undiscussed important topics like medicine. (Later in the chapter we shall be talking about what we would call biology.)

First there is *mathematics*. This goes back to the sixth century BCE and was primarily geometrical (Heath 1963). A major if shadowy figure was Pythagoras, born on the island of Samos around 570 BCE and founder of a secret cult with various rules for initiates, the most famous (or notorious) being that followers should abstain from beans! (This is perhaps not quite as silly as it sounds, for in fact many Mediterranean people are allergic to fava beans.) He and his followers were fanatics about mathematics, much of it a kind of mystical numerology, as they saw key ratios and connections linking different parts of reality. Pythagoras is best known, of course, for the famous theorem that bears his name. The Egyptians (as did the Chinese) knew that a 3, 4, 5 triangle is right-angled. It was Pythagoras or someone in his group who generalized it to all right-angled triangles (the square on the hypotenuse is equal to the sum of the squares on the other two sides) and who offered the geometric proof. Again, it was probably a follower who, through this and like theorems, was led to the investigation of irrational numbers, that is, those numbers that cannot be expressed as a fraction of two whole numbers. Specifically, the right-angled triangle formed by slicing a square diagonally has a hypotenuse that is irrational, being $\sqrt{2}$ (assuming the other sides to be 1, 1), a number impossible to express as m/n where m and n are whole numbers.

These and like ideas were formalized by Euclid two hundred years later, in his *Elements*. Here we have the first full example of an axiomatized system, with definitions – a point is that which has no part; a line is length without breadth, and so forth – and then axioms and postulates, truths assumed within the system, and theorems, things proven within the system. Today we tend to run axioms and postulates together, but for Euclid the axioms were the more geometrical truths – that a straight line can be drawn from a point to any other point, that all right angles are equal, and the like – and the postulates more general or logical truths – that if A is equal to B, and B is equal to C, then A is equal to C; and that the whole is greater than the part. With these very simple and obvious ideas – it is true that there was always debate about the fifth axiom (that parallel lines never meet) and that denying this in the nineteenth century led to non-Euclidean geometries – Euclid was able to develop and prove mathematical equations of great complexity and very much lacking in intuitive obviousness. Or rather, to give credit to his predecessors, Euclid was able to formalize and systematize many results that two or three centuries of hard labor had produced. He also gave a basis for the work of his successors, most notably Archimedes, known not only for his principle about the displacement of liquids but also for many other mathematical truths, often of great importance in technology and in developing theories of physics.

One topic of great interest to Euclid was that of the so-called Platonic solids. These are the five convex, regular polyhedra, that is, the solids that have faces, edges, and angles all congruent (equal). There is the tetrahedron with four equilateral triangles as faces, the hexahedron (cube) with six squares as faces, the octahedron with eight equilateral triangles as faces, the dodecahedron with twelve pentagons as faces, and the icosahedron with twenty equilateral triangles as faces. Euclid proved that there can be no more such

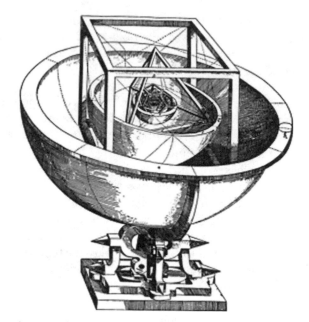

Figure 1.1. There are five and only five perfect solids – that is, solids with identical faces that are in turn regular polygons. In the *Mysterium Cosmographicum* (1596) Kepler (fancifully) suggested that by setting the solids with a common center (concentrically) in an appropriate order one could find the orbits of the planets. For him, a deeply committed Pythagorean-Platonist, all of reality was mathematical.

solids, and he spent much time in the *Elements* exploring their properties. He was not the first to discover such solids – there are models in prehistoric settlements, including those of the Neolithic people of Scotland around 1500 BCE. But it was Euclid who sealed their almost mystical significance, something that kept reappearing through history, most famously when in the sixteenth century the astronomer Johannes Kepler tried to find a formula for the distances of the planets from the sun by fixing distances according to the ways in which the solids could be fitted together concentrically [Figure 1.1].

It was what one might call the elegance, the beauty, of the solids that attracted the great philosopher Plato (428–348 BCE). His thinking can only be described as saturated with respect for and veneration of mathematics. It is clear that he was deeply influenced by the school of Pythagoras, to the extent that, in his great dialogue the *Republic*, the training of his leaders (the Rulers or Guardians) is modeled on the kind of society supposedly surrounding Pythagoras (Cooper 1997). As is well known, Plato distinguished between this world of ours in which we live and the Real World, the world of absolute being, the world of the Forms, capped above all by the Form of the Good. This was Plato's solution to a debate that had divided earlier Greek philosophers, notably Heraclitus, who argued that everything is in a state of flux ("You cannot step into the same river twice") and Parmenides, who argued that nothing moves or changes. ("How could what is perish? How could it have come to be? For if it came into being, it is not; nor is it if ever it is going to be. Thus coming into being is extinguished, and destruction unknown.") (Curd and McKirahan 1996) Plato's solution was that the world of Forms is unchanging, while our world is the world of change, of growth and decay.

Mathematics shows beauty and perfection. Although there is ambiguity about its exact status, its eternal truths belong to the world of Forms. This world of ours is in some way modeled on, a reflection of, the world of Forms. Hence, in Pythagorean style, we look to see mathematical ratios and equations ruling or governing the connections between things in our world. Inasmuch as they do – as in the harmonies we find in music or the ratios that govern great architecture – things are good, and inasmuch as they do not, things are bad or imperfect. The circle in particular has a special status, being thought of as the perfect figure, since it is equidistant from a point, and it is from here on that we find the obsession with explaining the heavens – thought of as being of far greater perfection than the world down here in which we live – in terms of uniform circular motion. To zig-zag around would not be dignified, nor would be acceleration and retardation.

Moving now to our second topic, although (as we shall see in a moment) far from abandoning mathematics, Plato (like his predecessors) was intensely interested in what today would fall under the heading of *chemistry*, the composition of this world of ours. He combined earlier beliefs that the materials of this world are made of small particles or corpuscles ("atoms") and that there are four basic elements: earth, water, air, and fire. In his great dialogue the *Timaeus*, of which we shall hear much more later (generally, although not universally, thought to be one of the later dialogues), Plato identified the corpuscles of earth, water, air, and fire with the shapes of the regular solids – cube, isahedron, octahedron, and tetrahedron, respectively. In today's language, this was a highly reductive system, because not only does it mean that mathematics is the secret to the underlying nature of this world of ours, it also means that the elements themselves are not basic, but decomposable into smaller units. These are smaller mathematical units of the same ultimate material and (cube excepted) with the same triangular faces and thus, with reshuffling, capable of moving from one substance to another. The physical properties of the units – fire has the sharpest points, earth the most solid and nonvolatile shape – explain the ways in which the elements strike us through our senses.

Aristotle (384–322 BCE), Plato's great student, seems at times to have formed his philosophy simply by putting negatives in front of all of his teacher's main verbs – a practice not entirely unknown today. In fact, although Aristotle did spend time criticizing Plato – for instance, he denied atomism (which has the particles moving in a void, something Aristotle thought impossible) – much more positively he was a deeply innovative thinker in his own right. This does not mean he was unwilling to borrow. This is especially true of his thinking about *astronomy* – our final topic for discussion here – for he based his thinking here on a theory of Eudoxus of Cnius, another of Plato's students (Kuhn 1957). Despite years of correction, even now people often think that, until the Scientific Revolution started when Nicholas Copernicus put the sun at the center of the universe, the general worldview was that the earth is flat and that folk will fall off if they get too close to the edge. Columbus, supposedly, was taking a big risk. This is total nonsense, an invention of the nineteenth century intended to make the medieval period – a time when Catholicism ruled the Western world – look absurd. The ancient Greeks were fully aware that the earth is a globe – apart from anything else, given the status of the circle, for someone mathematics–obsessed like Plato this was almost a necessity. But they were also persuaded by the obvious evidence, namely, by the eclipses of the moon when the earth's curved shadow crosses the lunar surface, and by the disappearance of ships as they cross the horizon – ships that stay visible longer from the tops of cliffs. Eudoxus's genius was to put an outer sphere, concentric with the earth, at the edge of the universe, always spinning on a north-south axis, with regular motion and carrying the stars around the earth approximately once a day.

At one stroke, one had the illusion of the stars moving across the heavens, as they do each night. Of course, he could not stop there – some explanation had to be given of the sun and the moon, not to mention the five wanderers, better known as the planets. Ingeniously, Eudoxus showed, and Aristotle endorsed, an onionlike picture, with the planets and the sun and moon being carried in invisible crystal spheres, likewise concentric, with their own regular motions around their own distinctive axes. Although the sun and moon show fairly regular motion through the heavens, the planets do not – every now and then they loop the loop (in technical terms, they "retrogress") – but by judiciously adding more and more crystal spheres (in total, you need twenty-eight), Eudoxus showed that you could get a qualitative illusion of planetary movement looking up from the earth. And the important point is that all of the motions are regular and circular – the kind of mathematical perfection one expects of the heavens [Figure 1.2].

To us, or rather to Aristotle, the important point is that this picture lays itself open to causal understanding. Why don't the planets go haywire or collapse down into the earth? Because they are held in place by the crystal spheres, and these keep them moving constantly. Why there is movement at all is something we shall discuss shortly. Nevertheless, causal though the theory may have been, professional astronomers – the people actually trying to quantify the motions of the heavens – did not find the theory of Eudoxus very satisfying or easy to work with. It had major empirical problems – most notably, it did not account for the brightness that planets show when they retrogress [Figure 1.3]. If, as everyone assumed, the planets are perfect and unchanging (being part of the heavens rather than of the imperfect earth), then they should not show changes, especially if they are always (taken individually) equidistant from the earth. The obvious explanation, that during retrogression they are brighter because they are closer, seemed to be barred. But not under a rival theory promoted a century later by Apollonius of Perga, who argued that the motions of the heavenly bodies, including the retrogressions, could better be mapped by the system of deferents and epicycles, where a planet goes round a circle (epicycle) the center of which is going around another circle (deferent). With this theory, one that lent itself at once to much more accurate quantitative results (given that one could fiddle around with radii of circles and, as need be, add epicycles to epicycles, treating circles both as deferents and as epicycles at

Figure 1.2. The geocentric world picture. From Peter Apian, *Cosmographia* (1524).

the same time), the brightness of the retrogressing planets comes out at once. They are closer then to the earth! [Figure 1.4].

It was this theory that was picked up and used by the greatest astronomer of antiquity, Claudius Ptolemy (83–161 CE), who lived and worked in Alexandria in Egypt. (For this reason, he is known as Hellenistic, meaning in the Greek tradition, rather than Hellenic, meaning Greek.) His major work on astronomy, the *Almagest*, was the definitive work on astronomy, holding sway for fifteen hundred years, right down to the time of Copernicus. However, note that what was gained on the roundabout was lost on the swings. The theory of deferents and epicycles was a wonderful tool for professional

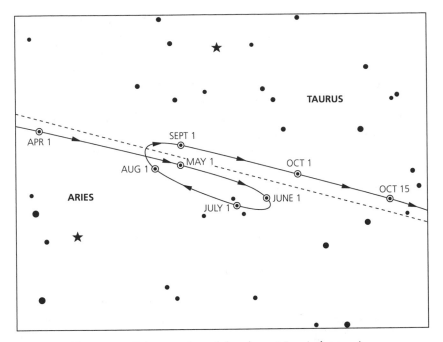

Figure 1.3. Retrogression of the planet Mars in late spring.

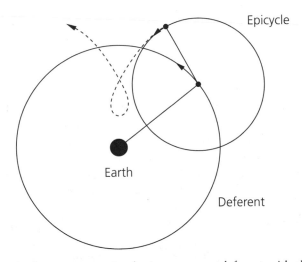

Figure 1.4. A planet on an epicycle, in turn on a deferent with the earth at the center, showing how you can explain the brightness of the planet during retrogression because this is the point at which it is closest to the earth.

astronomers, but at the cost of causal understanding. One could still keep in line with the Platonic insistence on mathematics, and the idea that perfection means above all that the circle dominates the structure and order of the world. However, no longer did one have a theory that made sense given Aristotle's physics. What keeps the planets moving, for instance? It cannot be circling crystal spheres, because these would have to cut across each other. Crystal spheres work only if they are concentric. This has led to much speculation about whether the ancient astronomers were what we would call instrumentalists – not really convinced that their theories truly described reality, but treating them as fictions that enable one to make accurate predictions. This is not really our problem, but we must keep in mind the tension between Aristotle's causal picture and the mathematical tools of the professional astronomer.

METAPHOR

I want now to pull back for a moment and to think about the nature of Greek science. And to do this, I want first to think about the nature of science generally. I suspect that most people if asked about the nature and purpose of science would say something like: "The human enterprise that describes and explains nature, the empirical world." If pushed, they would probably agree with the American sports commentator (the late Howard Cossell) that science aims to "tell it like it is." Well, science may aim this way, but in reality it does nothing of the sort. Or at least, it certainly does not achieve what it aims to do. For a start, we push onto the empirical world our picture of the way we think things ought to be. Take Plato and his atomism. He did not discover that the atoms are the perfect solids. He ordered, he hypothesized, that this is the way that they are. Note that I am not saying he was being stupid or unscientific. He had good theoretical reasons to think as he did, and once he had made the leap, he could tie in the atoms with empirical experience. Why does it hurt when we put our hand in the fire but not when we plunge it into the earth? Nor am I saying that things can never be changed. There was no way that Plato was going to change from seeing mathematics as all-important and finding circles everywhere, but the scientific community did change eventually – Johannes Kepler made the planets go around the sun in ellipses, not circles. What I am saying is that science is a dance, an interplay between the given and the found, the subjective and the objective, the mind-directing and the directed-mind.

This softens the way for an even more important point. Scientists try to build models; they try to provide pictures of the way things might be. Then they go out to see how their models fare when they are tested against the empirical world. To do this, they do above all what comes naturally in human thinking, they provide and use *metaphors* (Lakoff and Johnson 1980; Johnson 1981). Aristotle again: "Metaphor consists in giving the thing the name that belongs to something else." Why would we do this? Because in some sense "a good metaphor implies an intuitive perception of the similarity in dissimilar." But why care? Why bother? Because this leads to new ideas, new ways of thinking, to discoveries. "We will begin by saying that we all naturally find it agreeable to get hold of new ideas easily: words express ideas, and therefore those words are the most agreeable that enable us to get hold of new ideas. Now strange words simply puzzle us; ordinary words only convey what we know already; it is from metaphor that we can best get hold of something fresh. When the poet calls old age 'a withered stalk', he conveys a new idea, a new fact, to us by means of the general notion of 'lost bloom', which is common to both things" (*Rhetoric* 1410b10–15, Barnes 1984, 2250).

Metaphors stimulate us to think of old things in new ways. Literally speaking, they are false. Plants have stalks, not humans. But they push us in new directions. As the philosopher Max Black (1962) used to say – using a metaphor to talk about metaphor – we view objects through the lens of the metaphor. We view humans through the lens of a faded flower, as something once beautiful and vibrant but now old and coming to an end. Not as something always worthless, but as something beyond the point of full flourishing and pointing to death. Science uses metaphor – natural selection, continental drift, force, work, attraction, charm, genetic code, Oedipus complex – to structure experience and to build our models of understanding. Nature does not literally select; magnets do not literally attract one another; the genes are not truly written in code – but it suits us to think of them in these ways, because then we can make discoveries. We find that some butterfly wings are used for camouflage because those putative ancestors that escaped detection survived and reproduced, and those that did not did not. We find that a magnet can be used to tell direction because it swings to an angle pointing to the poles. We find that this gene causes brown hair and that gene lies behind blond hair. Without our metaphors we are blind – metaphorically, at least. But the use of metaphor again highlights the fact that we put our own ideas into the product and do not just read information off virgin experience.

There has been much written about whether or not metaphors are essential for human thought. Aristotle may have thought not. Later thinkers have certainly thought not. Thomas Hobbes, the seventeenth-century English philosopher, absolutely thought not. "For though it be lawful to say (for example) in common speech, *the way goeth, or leadeth hither or thither; the proverb says this or that*, whereas ways cannot go, nor proverbs speak; yet in reckoning, and seeking of truth, such speeches are not to be admitted" (Hobbes 1998, 31). Today's philosopher Jerry Fodor agrees: "When you actually start to do the science, the metaphors drop out and the statistics take over" (Fodor 1996, 20). One can see why some people are tense about metaphors. They are false! Oh yes, you do have similarities, but when push comes to shove they are not true, and thinking that they are leads to all sorts of mistakes. We wonder if electricity will leak out when a copper pipe is pierced or if people with red hair – fiery – are hot-tempered. However, they do lead to insights, and others argue the other way, claiming that you simply cannot get rid of metaphor. They are a necessary component to human thinking.

I confess that although my inclination is to say that metaphors are necessary, I am not quite sure how one would prove this definitively. I would agree, however, that they are used throughout science, and I cannot imagine science without them. They certainly seem indispensible in discovery – they have great heuristic use – and thus for all of the dangers seem here to stay. By way of example, consider the metaphor of a "division of labor" (Ruse 1999b). This idea was introduced by Adam Smith, the eighteenth-century Scottish philosopher and political economist. If you are making pins, you do much better if you give different people different jobs rather than trying to have everyone do everything. The biologists picked it up in the nineteenth century, notably the Belgian-born Henri Milne-Edwards, who spoke of a physiological division of labour, with different parts of the body doing different things. Darwin grabbed it and used it repeatedly:

The advantage of diversification in the inhabitants of the same region is, in fact, the same as that of the physiological division of labour in the organs of the same individual body – a subject so well elucidated by Milne Edwards. No physiologist doubts that a stomach by being adapted to digest vegetable matter alone, or flesh alone, draws most nutriment from these substances. So in the general economy of any land, the more widely and perfectly the animals and plants are diversified for different habits of life, so will a greater number of individuals be capable of their supporting themselves. A set of animals, with their organisation but little diversified, could hardly compete with a set more perfectly diversified in structure. (Darwin 1859, 115–16)

It is used, too, by modern biologists. The Harvard expert on the social insects, Edward O. Wilson, has done fine-grained analyses of the caste structure in some of the Brazilian leaf-cutter ants, the *Atta* genus (Wilson 1980a, b). They are remarkable insects, with soldiers and foragers and leaf cutters and (back in the nest, where they grow a fungus to feed to their young on the chewed-up leaves) gardeners, and gatherers, and nurses – not to mention the young, the queen, and the idle males. Trying to work out the ratios of the various forms, Wilson relies repeatedly on the division-of-labor metaphor: what is the most efficient way to use resources in the group, and how has selection achieved this? "The fungus-growing ants of the tribe Attini are of exceptional interest because, to cite the familiar metaphor, they alone among the ants have achieved the transition from a hunter-gatherer to an agricultural existence." Among females, there are no less than seven castes of workers, plus queen and larvae ("which may serve some as yet unknown trophic function").

A key feature of *Atta* social life disclosed by these data is the close association of both polymorphism and polyethism with the utilization of fresh vegetation in fungus gardening.... An additional but closely related major feature is the 'assembly-line' processing of the vegetation, in which the medias cut the vegetation and then one group of ever smaller workers after another takes the material through a complete processing until, in the form of 2-mm-wide fragments of thoroughly chewed particles, it is inserted into the garden and sown with hyphae.... (Wilson 1980a, 150)

Thinking now in terms of how natural selection might have achieved all of this in an optimal fashion – maximum amount of output for a minimal amount of input – we see that the ants use the most efficient ways of dividing the tasks. "What *A. sexdens* has done is to commit the size classes that are energetically the most efficient, by both the criterion of the cost of construction of new workers . . . and the criterion of the cost of maintenance of workers."

Accept now the significance of metaphor in science. In some sense, some metaphors are going to be more important, more basic than others. These are what one might call root metaphors. I don't necessarily want to define this term in such a way as to say that there can be one and only one root metaphor at any one time. What I do want to say is that root metaphors are somehow very basic, and other metaphors fit in with or derive from them. By way of illustration, stay for a moment longer with Darwinian evolutionary biology. The one thing that Charles Darwin does above all other things in the *Origin of Species* is explain (try to explain, if you insist) the designlike nature of the living world (Ruse 2003). Hands, eyes, teeth, noses, leaves, bark, roots, are all as if designed. They work to the ends of their possessors. Because I have hands and eyes and teeth and so forth, I can better succeed in life's struggles. I am more likely to survive and reproduce. I am more likely to be naturally selected. This then brings us face to face with what I would call a fundamental root metaphor in Darwinian evolutionary biology – the underlying metaphor that makes sense of the subject. It is the metaphor of the world of organisms as designed. It may be that they were literally designed by God, but that is not the point in Darwinian biology. They are *as if* designed – truly not designed, but appearing so because of natural selection. Everything else is on top of – or, if you prefer, embedded within – this metaphor. This is true of Wilson's discussion of the leaf cutters. They show a division of labor, but this is an example of great design. The same is true of other metaphors that evolutionists use. Take the popular one of an arms race. We think of competing lines of organisms – prey and predator – as engaged in competition like that we find between human nations.

> The evolution of the ungulates is not adapted merely to greater efficiency in securing and digesting grass and leaves. It did not take place in a biological vacuum, but in a world inhabited, *inter alia*, by carnivores. Accordingly, a large part of ungulate adaptation is relative to the fact of carnivorous enemies. This applies to their speed, and, in the case of the ruminants, to the elaborate arrangements for chewing the cud, permitting the food to be bolted in haste and

chewed at leisure in safety. The relation between predator and prey in evolution is somewhat like that between methods of attack and defence in the evolution of war. (Huxley 1942, 495–6)

Notice that, as with the division of labor, underlying this is the notion of design. Battleships are designed to withstand attack – heavy amor, for instance – and guns are designed to attack – penetrating power, for instance. Likewise in the animal world – ungulates are designed to gulp down their food and digest later, and carnivores for speed in chase.

THE ORGANIC METAPHOR: PLATO

Turn back now to the Greeks. What is the ultimate root metaphor operating in their science? It is the idea of the world as living, the world (the earth, at least) as an organism (Sedley 2008). The world is not dead matter, but something organic, with life. Start with Plato. In today's language, he was a "teleologist." He thought that all things, what we would call both living and dead – the once alive and the never alive – show ends. They exhibit purposes. It makes sense to ask: What are they for? Plato's was what we today call an external teleology. Things show ends because someone, not part of (external to) the system, designed them. All of this comes through very clearly in the *Phaedo*, the dialogue that Plato wrote about Socrates' last day on earth. Think of why someone grows. "I had formerly thought that it was clear to everyone that he grew through eating and drinking; that when, through food, new flesh and bones came into being to supplement the old, and thus in the same way each kind of thing way supplemented by new substances proper to it, only then did the mass which was small become large, and in the same way the small man big" (96d). But then Plato argues that this will not do on its own as an explanation. It is incomplete. We need value and purpose. "The ordering Mind ordered everything and placed each thing severally as best that it should be, so if anyone wanted to discover the cause of anything, how it came into being or perished or existed, he simply needed to discover what kind of existence was *best* for it, or what it was best it should do or have done to it" (97b-c). In other words, a simple explanation in terms of how things act is not enough. "If someone had said that without bones and sinews and all such things, I should not have been able to do what I decided, he would be right, but surely to say that they are the cause of what I do, and not that I have chosen the best course, even though I act with my mind, is to speak very lazily and carelessly" (99a-b).

Now we turn again to the *Timaeus*, by far the most influential of Plato's dialogues for nigh two thousand years. Here we find a creation story. It is not the Judeo-Christian story of God creating everything out of nothing by an act of will, nor yet of a God who decided absolutely how things should be. It is a story of a god, or a craftsman (the "Demiurge"), who ordered or designed an already-existing world, and who was guided in his work by the template shown in the world of Forms. "Now surely it's clear to all that it was the eternal model that he looked at, for, of all the things that have come to be, our universe is the most beautiful, and of causes the craftsman is the most excellent. This, then, is how it has come to be: it is a work of craft, modeled after that which is changeless and is grasped by a rational account, that is, by wisdom" (Cooper 1997, 1235).

Next comes the crucial bit:

> Now it wasn't permitted (nor is it now) that one who is supremely good should do anything but what is best. Accordingly, the god reasoned and concluded that in the realm of things that are naturally visible no unintelligent thing could be as a whole better than anything which does possess intelligence as a whole, and he further concluded that it is impossible for anything to come to possess intelligence apart from soul. Guided by this reasoning he put intelligence in soul, and soul in body and so he constructed the universe.

Notice that, for all that Greek thought did have major influences on Christian theology, for Plato the soul is not straightforwardly identical to the Christian notion of something that is our inner essence, somehow connected to thinking, and by which we are made in God's image. In Augustine's words, "a special substance, endowed with reason, adapted to rule the body" (*De Animae Quantitate* 13.22; *De Genesi contra Manicheos* II.11); in the words of the Catholic catechism, "the innermost aspect of man, that which is of greatest value in him, that by which he is most especially in God's image." In other words, the "spiritual principle" in humans. For Plato, soul is more the life force that animates beings. It is not necessarily thinking, but can be. In the *Republic*, the soul has three parts – intellectual, spirited, and appetitive. So what Plato is saying here is that, in the case of the earth, we have a living being, one with soul. It is also incidentally a living being that is an intelligent living being.

But why just one such being? Why is the world not filled with living beings?

> When the maker made our world, what living thing did he make it resemble? Let us not stoop to think it was any of those that have the natural character of a part, for nothing that is a likeness of anything incomplete could ever turn out beautiful. Rather, let us lay it down that the universe resembles more closely than anything else the Living Thing [that is the Form of Life] of which all other living things are parts, both individually and by kinds. For that Living Thing comprehends within itself all intelligible living things, just as our world is made up of us and all the other visible creatures. Since the god wanted nothing more than to make the world like the best of the intelligible things, complete in every way, he made it a single visible living thing, which contains within itself all the living things whose nature it is to share its kind. (Cooper 1997, 1236)

The world is alive. We expect to see changes and flowings to and fro – which, of course, with the seasons and the rivers and the seas and all else, we do see. It makes sense to ask about parts and how they function in the whole. Why do we have things? Clearly rainfall is for our benefit, and the warmth of the sun likewise. And notice above all that things have value – real, objective value. The rainfall is something of value to the world, just as our fingernails are of value to us. The Craftsman, the Demiurge, did not do things for fun or by chance. He did them because they were for the best. He wanted to imitate the world of Forms as much as possible. Hence the mathematics underlying the world's structure and the worth of the functioning parts of the world, parts to be understood only in relationship to the whole.

THE ORGANIC METAPHOR: ARISTOTLE

Turn now to Aristotle. He too was a teleologist, but of a rather different kind. Start with causation. For Aristotle, this is the heart of scientific understanding. "Knowledge is the object of our inquiry, and men do not think they know a thing until they have grasped the 'why' of it (which is to grasp its primary cause)" (*Physics* 194b18–20,

Barnes 1984, 332). Famously, he distinguished four kinds of cause. The first is what we call "material" cause, the stuff from which something is made: the marble from which a statue or sculpture is made. Then there is "formal" cause, the pattern or archetype behind the creation: in the case of the *Pieta*, it would be the idea of the dead Christ. The "efficient" cause is the maker: Michelangelo. And finally the "final" cause, the end or reason for the making: to glorify God, not to mention the French cardinal who commissioned it. It is final cause that gives the teleology to things in the world. This is often referred to as an internal teleology as opposed to Plato's external teleology; Aristotle does not postulate a designer in the way of Plato. Rather, there seems to be something in the nature of things themselves – internal to things – that involves organization or intention or purpose. This is not necessarily intelligent, although of course it can be, but rather a kind of vital force, one that leads to organization.

Most obviously we see this in organisms. (Aristotle had spent much time studying organisms, to the extent that without too much anachronism one can almost describe him as a professional biologist.) They have a kind of teleology in their very reproduction and continued existence, from one generation to the next. "If then it is both by nature and for an end that the swallow makes its nest and the spider its web, and plants grow leaves for the sake of the fruit and send their roots down (not up) for the sake of nourishment, it is plain that this kind of cause is operative in things which come to be and are by nature" (*Physics* 199a25–30, Barnes 1984, 340). At times, this can sound almost Platonic, because nature seems personified. "It is plain then that nature is a cause, a cause that operates for a purpose" (*Physics* 199b32, Barnes 1984, 341). But nature is not really a thinking being, but rather something that has organization and exhibits final cause. We see this clearly when we turn specifically to organisms taken individually, as we try to understand the parts and the reasons for their existence. Here the discussion becomes explicitly functional, appealing to final causes. Consider the discussion of why we have hands.

> Standing...erect, man has no need of legs in front, and in their stead has been endowed by nature with arms and hands. Now it is the opinion of Anaxagoras that the possession of these hands is the cause of man being of all animals the most intelligent. But it is more

rational to suppose that man has hands because of his superior intelligence. For the hands are instruments, and the invariable plan of nature in distributing the organs is to give each to such animal as can make use of it; nature acting in this matter as any prudent man would do. For it is a better plan to take a person who is already a flute-player and give him a flute, than to take one who possess a flute and teach him the art of flute-playing. (*Parts of Animals* 687a5–14, Barnes 1984, 1071–2)

We have hands to aid the use of our intelligence. In certain respects, however, at least as important is the way in which Aristotle continues his argument. We are dealing not just with physical body parts here but with the whole way in which the world is constituted and how it should be understood.

For nature adds that which is less to that which is greater and more important, and not that which is more valuable and greater to that which is less. Seeing then that such is the better course, and seeing also that of what is possible nature invariably brings about the best, we must conclude that man does not owe his superior intelligence to his hands, but his hands to his superior intelligence. (*Parts of Animals* 687a14–19, Barnes 1984, 1072)

The world – nature – is not a blind system, but something (even if not by direct intelligence) that is striving to get things right, to make things as good as possible. The world in which we live is one with intrinsic value, worth – just as is Plato's – and like Plato's world it is one where "soul" in some sense plays a crucial role. As in Plato, it is some kind of animating force, what we might call a "vital" force. In the *De Anima*, the work devoted to this topic, Aristotle distinguishes the mere vegetative force from the animal force and the intellectual force – this last possessed only by *Homo sapiens*. Although Aristotle is denying the external teleology of Plato, it is crucial to see that this notion of force is bound up with Aristotle's views on the working of the world, which are conceived in terms of ends, of final causes. Although he denies the atomism embraced by Plato, Aristotle is equally committed to the four substances – earth, water, air, and fire. These are crucial to his physics because, in contrast to the later physics of Newton – where, thanks to the first law of motion, we expect moving bodies to keep going under their own momentum – Aristotle thought that if there is no force acting directly on something, it would simply stop moving or functioning. Which raises the question of why, if we release a ball from a clenched fist, it falls to the ground. Why, if we open an empty bottle under the water, do the bubbles rise to the top? Aristotle thought that all substances have their natural places, their right places, and in some wise they are forever striving to get to them. Things move in order to find their proper places. Earth at the center of the universe, then water, then air, and finally fire at the top. The movement of composite bodies is "determined by that simple body which prevails in the composition. From this it is clear that there is in nature some bodily substance other than the formations we know, prior to them all and more divine than they. Or again, we may take it that all movement is either natural or unnatural, and that the movement which is unnatural to one body is natural to another – as for instance is the case with upward and downward movements, which are natural and unnatural to fire and earth respectively" (*On the Heavens* 269a29–269b1, Barnes 1984, 449).

We may not have quite the world spirit of Plato, but we have a world that is living, that is certainly not dead and lifeless. And to this you can add, as a kind of crescendo or apotheosis, Aristotle's views on what keeps the spheres of the planets moving along – for given his physics, there must be something. Aristotle is way too sophisticated a thinker to accept the traditional Greek gods, but there is a point to this kind of mythologizing. There has to be something keeping the heavens in action: "this is plain not only in theory but in fact. Therefore the first heavens must be eternal. There is therefore also something which moves them. And since that which moves and is moved is intermediate, there is a mover which moves without being moved, eternal, substance, and actuality" (*Metaphysics* 1071a23–26, Barnes 1984, 1694). These are Aristotle's famous – or notorious – Unmoved Movers. They are divine, perfect, in some sense the cause of the world and in some sense the purpose. Cause not so much in the sense of physically creating things – for Aristotle, things were eternal, not just in substance but also (against Plato) in form – but in the sense of responsible for things, and hence in another way the reason or end purpose of things.

In other words, even though Aristotle did not have Plato's straightforward picture of the world in which we live as a created organism, in major respects his thinking was much along the same lines. The underlying root metaphor that guided his thinking, the metaphor that functioned as a lens through which he viewed and tried to understand the physical world, was that of an organism. To understand things, we must ask about ends – about final causes – and this gives both a heuristic and a sense of value. When things are working properly, this is good; and when not, this is bad. Such then was the Greek view of nature.

TWO

THE WORLD AS A MACHINE

No Christian could ultimately escape the implications of the fact that Aristotle's cosmos knew no Jehovah. Christianity taught him to see it as a divine artifact, rather than as a self-contained organism. The universe was subject to God's laws; its regularities and harmonies were divinely planned, its uniformity was a result of providential design. The ultimate mystery resided in God rather than in Nature, which could thus, by successive steps, be seen not as a self-sufficient Whole, but as a divinely organized machine in which was transacted the unique drama of the Fall and Redemption. If an omnipresent God was all spirit, it was the more easy to think of the physical universe as all matter; the intelligences, spirits and Forms of Aristotle were first debased, and then abandoned as unnecessary in a universe which contained nothing but God, human souls and matter.

Hall 1954, xvi–xvii

There is much that is relevant to our inquiry in this introductory paragraph, drawn from one of the great classics on the history of the Scientific Revolution – the period in the sixteenth and seventeenth centuries that takes us from Nicholas Copernicus announcing that the earth moves around the sun to the great Isaac Newton bringing all together under his law of universal gravitational attraction. For a start, most obviously, we have the move from the world-as-organism metaphor to the world-as-machine metaphor. For a second, no less important for our inquiry, we see that the move was not one fueled by a desire to deny or belittle the Christian God. In major respects, it occurred because of belief in the Christian God. For a third, we start to see how it is that this nevertheless held the seeds of the expulsion of God from the cosmos – with the possibility of denying His existence, certainly His pertinence, altogether.

PLATO, ARISTOTLE, AND PLATO AGAIN

But to pick up our story in the right place, we go to the Middle Ages, especially to the so-called High Middle Ages, beginning around the eleventh century. Everybody knows – more precisely, everybody "knows" – that for science, after the Greeks and under the Romans, it was mainly a matter of cleaning up and elaborating. Rather than disinterested inquiry for its own sake, the latter were more into empire building and the technology that goes with it – roads and viaducts and killing machines and the like. Then, with the fall of Rome came the Dark Ages, and as far as science was concerned, that was it for a millennium. It was not until the sixteenth century that things started to pick up again.

There is some truth in this story, but much is exaggeration – an exaggeration by those in the Scientific Revolution wanting to stress their own originality (Lindberg 1992, 1995; Grant 1996). Certainly, there were long periods of drought. But this was in the West. The coming of Islam, starting in the seventh century – Muhammad lived from around 570 to 632 CE – saw the rise of a great civilization, one that cherished and furthered the cause of science. Particularly in areas like astronomy and medicine the Arabs made very great strides. Then in the West, in part stimulated by contact with Islam (both in the East and in the Iberian peninsula, for much of the time under Arab rule), the four centuries of the High and Late Middle Ages saw much interest in science and significant moves forward, albeit always within the framework of Aristotle's thinking. Best known and very important was the devising and developing of the notion of impetus. A major query raised by Aristotle's physics is why, before ultimately falling to the ground, something like a javelin travels a distance when thrown by an athlete. Being essentially earth, why does it not fall directly to the ground on being released? All kinds of ad hoc hypotheses had to be invented, suggesting that somehow the surrounding air was excited and thus the javelin traveled sideways before falling. For Aristotle, it was essential that the air keep working as a cause; otherwise, the javelin would drop immediately.

We of course would argue that the philosopher had got things backward – for us, the javelin has momentum, and the query is why it does not simply keep moving forward once it is released. We argue that it falls because now a new force acts upon it, namely, earth's gravity. Some proto-ideas about momentum go back to the sixth century and the Alexandrian philosopher John Philoponus, but it was not until the fourteenth century that Jean Buridan (1300–1358), a French cleric, developed the idea of impetus, a kind of force that is imparted to a body when it is thrown – close to momentum but a little different in that it was supposedly something working actively to keep the body moving rather than something that the body had and that could be dissipated only through friction and collision and the like. This proved a very fruitful idea and lent itself to mathematical treatment, perhaps a clue to why, when the Scientific Revolution did come, it was physics more than anything else that sprang ahead in development and toward the goal of scientific maturity.

What about the philosophers, and what about the metaphor of the world as an organism? Plato and Aristotle dominated the scene. First, in the twelfth century, came Plato – although basically only the Plato of the *Timaeus*, and of that, only the first part dealing with the cosmos. (The second part deals with humans.) Then, in the thirteenth century, Aristotle was rediscovered, and of course became the basis for the great philosophical system woven by Saint Thomas Aquinas. (Interestingly, although perhaps predictably, given that the original sources were only then being uncovered, Thomas could not read Greek and had to rely on the Latin translations of others.) Then, by the fifteenth century, we get a rise again of Platonism, this time of the full Plato, especially of works like the *Republic*. Through all of these phases the organic metaphor rode secure, even though the seeds of change were being sowed, if only because of the above-mentioned problem of reconciling the Greeks' thinking about the universe with the Christian story, particularly as taken over from the Ancient Jews.

Above all the *Timaeus* was the text for the idea of the world as an organism, and that was absolutely the root metaphor of the High Middle Ages. It had to be agreed that Plato was wrong in thinking the universe eternal and equally wrong in supposing that the Demiurge simply ordered the preexisting universe rather than creating it from nothing. But after this, things seem to fit pretty well. Admittedly the doctrine of the Trinity needed a little fancy intellectual footwork, but there were ways of tackling this. John of Salisbury (1120–1180), a student of Peter Abelard, wrote:

> ...in the books of Plato we find many things consonant with the words of the prophets. In the *Timaeus*, for example, during his subtle investigation of the causes of the world, he seems clearly to express the Trinity which is God, when he locates the efficient cause in the power of God, the formal cause in his wisdom, and the final cause in his goodness, which alone induced him to make all creatures sharers in his goodness, according as the nature of each one is capable of beatitude. At the same time he seemed to understand and teach that there was a single substance in these, when he asserted that the craftsman and the shaper of the world was one God, whom, because of his extraordinary goodness and sweetness, he called the Father of all, and whom, because of the boundlessness of his majesty, power, wisdom, and goodness, it is as difficult for us to find as it would be impossible for us to proclaim, should we ever find him. (Dronke 1988, 56–7)

Although there was not complete agreement on this, God was the creator/designer; Jesus was the wisdom, as represented by the Forms; and the Holy Ghost was the world spirit or soul. (You have to be careful here not to indentify God with the world, with His creation. That is the heresy of pantheism.)

> ...the adornment of the world is whatever can be seen in the individual elements, such as the stars in heaven, the birds in the sky, the fish in the water, men on the earth, etc. But because some of them are always in motion, some grow, some discern, and some feel, and they have this property not from the nature of the body but from the nature of the soul. Thus Plato begins by discussing the soul, that is, the world-soul. And the world-soul is a kind of spirit inherent in things, conferring motion and life on them. (Dronke 1988, 68)

Aristotle also incorporates this kind of thinking, and one sees it in his medieval followers. Take, for instance, Saint Thomas's fifth proof of the existence of God. It is explicitly teleological:

> The fifth way is taken from the governance of the world. We see that things which lack intelligence, such as natural bodies, act for

an end, and this is evident from their acting always, or nearly always, in the same way, so as to obtain the best result. Hence it is plain that not fortuitously, but designedly, do they achieve their end. Now whatever lacks intelligence cannot move towards an end, unless it be directed by some being endowed with knowledge and intelligence; as the arrow is shot to its mark by the archer. Therefore some intelligent being exists by whom all natural things are directed to their end; and this being we call God. (Aquinas, *Summa Theologiae* Ia, q. 2, a. 3)

Then, with the rise again of Platonism toward the end of the Middle Ages, the organic metaphor seems to become even more secure. It was not just something embraced by the learned and the philosophical; it was part of general folklore, too (Merchant 1980). Miners naturally thought of the metal "veins" running through the underground strata as evidence that we are dealing with a living body – one that could regenerate with the right treatment. Farmers and other rural folk thought the changing of the seasons clearly confirmed that we are denizens of a live being – one that grows, flourishes, and dies each year. Things like volcanoes, and lightning, and running streams and rivers all likewise pointed to the vibrant, living earth on which we live. We might now think (rightly) that we are dealing with a metaphor, but for the medievals this notion that the world is indeed living was something that needed no real proof. Plato and the other philosophers were simply pointing to the obvious.

COPERNICUS

I turn now to the growth of science from Copernicus to Newton. Remember, I am not telling a straight or simple history of science. Many have done this before and better than I. The aim is to trace the change of metaphor from organism to machine. The most celebrated theory of scientific change is that of the late Thomas Kuhn in his *The Structure of Scientific Revolutions*. He sees scientists working within paradigms, world pictures, and the change as being abrupt and total – now you don't have it, now you do. However, as Kuhn's own work on the history of science showed very clearly, things are much more complex than this. The greatest of revolutionaries often show significant debts to their past (for all that they often try to conceal them), until one wonders at times whether they are the last of the old or the first of the new. Certainly this is true of Nicholas Copernicus, whose *De Revolutionibus Orbium Coelestium* (*On the Revolutions of the Celestial Spheres*) was published in 1543 and supposedly handed to the author on his deathbed. He was a minor cleric (not ordained) and ostensibly a church administrator but in truth a professional astronomer. Copernicus therefore knew intimately the *Almagest* of Ptolemy and was simply unimpressed by what he took to be the clumsiness of the geocentric system. He therefore put the sun at the center, thus promoting a heliocentric system.

However, although he was a revolutionary in this respect – it is unclear just how much Copernicus knew of earlier attempts to make the sun central to the universe, for instance, that of Aristarchus of Samos (310–230 BCE) – in many other respects Copernicus stood in hallowed traditions. Most importantly, he never thought to question the circular motions of the heavens. More than this, he clearly kept to some kind of teleological view of the workings of nature. Things can no longer fall to the center of the earth simply because it is the center of the universe, but particles of matter on the various planets (including earth) have a natural tendency to move to the center of their particular domain.

One thing that Copernicus did not do was to provide a theory that was any more predictively accurate than that of Ptolemy. Nearly a millennium and a half after the *Almagest* was written, it was very clear that nothing was very exact – what might once have been roughly true was increasingly out of kilter. Copernicus did nothing to remedy any of this. For him, it is clear that the aesthetic issues were all-important. Pertinent to this, it seems probable that Copernicus felt and was influenced by the waves of Plato worship that he encountered when he left his native Poland and went to study in Italy. Plato, in turn influenced by the Pythagoreans, made much of the status of the sun – it is the source of light as well as the energy that makes possible our life here on earth. In the *Republic*, Plato likens the sun to the ultimate Form of the Good, something that Christian commentators identified with the Godhead. The Pythagoreans, in fact, thought that the Fire of Zeus is the center of the universe, that the earth goes around this and that we are illuminated by the sun, a kind of mirror reflecting the ultimate fire. Certainly, Copernicus was capable of florid rhetoric when it came to the sun. "He is rightly called the Lamp, the Mind, the Ruler of the Universe; Hermes Trismegistus [supposed writer of sacred texts from the time of Moses] calls him

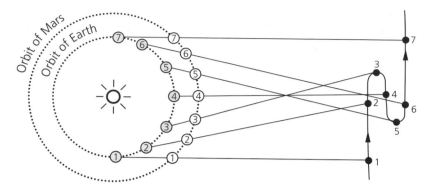

Figure 2.1. Copernicus's explanation of the regression of Mars. Note that this is the point at which Mars seems (to us) to be farthest from the sun.

the Visible God, Sophocles Electra calls him All-seeing. So the sun sits as upon a royal throne ruling his children the planets who circle round him" (Hall 1983, 70). More than this, Copernicus was obsessed with mathematics, something that Plato (and before him the Pythagoreans) had made much of. There is even the suspicion that he wrote in deliberately obscure terms, so that only initiates (that is, fellow professional astronomers) could follow his writings – again something in the Platonic-Pythagorean tradition.

Be these points as they may, for Copernicus it was the elegance of the heliocentric system that was all-important – the significance of the sun and the astronomical consequences of making the sun the center of the universe. Significant here was the way in which the heliocentric system had such a ready explanation for the two kinds of planets, the inferior ones (Mercury and Venus), which always stay close to the sun, and the superior ones (Mars, Saturn and Jupiter), which wander away from the sun and retrogress only when farthest away (in opposition) [Figure 2.1]. These facts follow from the inferior planets being between the earth and the sun and the superior planets being beyond the orbit of the earth and out toward the stars. Apparently not a great factor at this point – pro or con – were the possible religious implications. It is true that Copernicus's editor had added a preface that denied that one had to take the theory literally – it was just "saving the appearances" (as philosophers today speak of such instrumentalist theories) – but he himself probably thought it literally true and did not feel any great tension between the theory and his Christian beliefs. Although Copernicus's ideas leaked out – the

Pope heard with interest an exposition of his ideas – generally there was little religious opposition, and to be frank not a great deal of scientific interest. There were only about ten Copernicans in the whole of the sixteenth century, some Protestant and some Catholic (Westman 1986). The story that Luther sneered at Copernicus is apparently apocryphal.

As much as anything, of course, the big problem was how any of this was supposed to work. The traditional Aristotelian theory using concentric spheres made causal sense. Already the professional astronomers with their recourse to deferents and epicycles were making causal nonsense of this – and as a professional, Copernicus was as happy as his Ptolemaic contemporaries to use such devices to get more accurate results. Now, with the heliocentric hypothesis, the whole causal picture seems to fall apart. For a start, with the moon circling the earth, no longer the center of the universe, we seem to have something cutting through all kinds of heavenly spheres.

TYCHO, KEPLER, AND GALILEO

Why go down the Copernican route when in many respects nothing much seemed to hang on it, and when in other respects there seemed to be uncomfortable consequences? Most notably, there was the absence of stellar parallax, the kind of disturbing effect one should see in the positions of the stars if the earth were moving. It could be absent only if the stars were so far away that such parallax would be undetectable – far enough away that the diameter of the universe would have to be increased 400,000 times from the then-believed measure of 98 million miles. One who felt the attractions of the Copernican system and yet could not stomach the consequences was the Danish astronomer Tyco Brahe. He was spurred to offer a hybrid theory, putting the earth at the center, with the sun circulating, and everything else going around the sun. Formally, you could get the same results as from Copernicus – as from Ptolemy, for that matter – but you avoided moving the earth, and you stayed away from the problems of stellar parallax. But if anything was ugly, the very opposite of elegant, it was this theory. And now, of course, the causal issues are really desperate. Spheres are constantly cutting across spheres. How does anything stay in place?

A number of things were needed if Copernicus's insights – perhaps we should call them intuitions – were to be carried forward and developed. First, someone had to do a much better job of mapping and interpreting the movements of the heavens. Today, it is for the mapping that we celebrate Tyco rather than for his cosmological speculations. He is justly famous for having made a whole new survey of the heavens and having produced measured readings that were orders of magnitude better than anything produced formerly. Working from these, Brahe's student Johannes Kepler did the interpreting needed to carry forward the Copernican insights. An ardent heliocentrist, Kepler was an even more fanatical Pythagorean than Copernicus. He saw mathematical forms and figures and relationships everywhere, famously (or notoriously) putting the planetary orbs in order according to the ways in which they fit into concentric perfect figures, as mentioned in the previous chapter. Only someone with a very odd idea – that is to say, a very Platonic idea – of God would do something like that. More lastingly, it was Kepler who worked out fundamental regularities obeyed by the planets, including one that smashed two millennia of Greek assumptions about the necessity of starting with (and staying with) circular motion when analyzing the heavens. Kepler's first law of planetary motion stated that planets move in ellipses, with the sun at one focus. It is true that the motions are not that far from circular, but no longer could there be any pretense that heavenly motion was genuinely circular, or analyzable into real circles. Thanks to Brahe, Kepler realized that the planets have other paths.

The second thing needed was for someone to start working on causes. For over fifteen hundred years, people had been living in two worlds – the causal world of Aristotle and the saving-the-appearances world of the professional astronomers. Now things were getting worse and worse. More than this: some of the biggest factors – impediments, perhaps – were starting to crumble. Under the Aristotelian system, the division between our world and the world of the heavens is absolute. They are simply different systems with different rules. Up there, everything is uniform, circular, eternal – god-driven. Down here, it is messy, changing, variable, and everything else – moreover, the rules are different, with the elements determining how things move, up or down. With the earth demoted – changed, if you like – from its special place in the universe; with the universe now perhaps starting to seem infinite (with the stellar parallax issue making the universe that much bigger, it did not take long for some to push the boundaries out all the way); with orbits cutting across orbits; with the circular motion of the heavens under attack; and with everything else, it was no longer making much sense to insist on two kinds of physics, one up there and one down here.

The great Italian mechanist Galileo Galilei was the point man here, for it was he who picked up on the problems of bodies in motion, kinetics, and developed the field far beyond the primitive state in which he had found it. He was working both positively and negatively. The story of his dropping different-sized objects from the Leaning Tower of Pisa, to show that Aristotle was wrong in thinking that heavier bodies fall faster than lighter ones, is a good story. And like many good stories, it is probably untrue. If it did happen, it may even have been an opponent trying to prove Galileo wrong – try dropping a piece of balsa wood and a piece of oak and see what happens given the resistance of the air. But like many myths, there is truth here. Galileo was working out the principles of motion, and at the same time he was showing that the old Aristotelian system was incorrect. From our viewpoint, what he was doing particularly was showing that there is no need to appeal to final causes. Bodies can be understood in their own right, as objects at rest or moving, under forces that are placed upon them, rather than striving to achieve ends that nature desires. Efficient causes are what do the job.

This new physics was to prove essential in the final push to overthrow the old ways of thinking, although, perhaps not that surprisingly – a little bit like Moses, who led his followers to the Promised Land but could not enter himself – Galileo mimicked Copernicus in combining the old and the new. He may have provided the means to apply the same physics throughout the universe, but he himself still clung to the old ways, thinking that the motions of the heavens are circular – for all that he was in correspondence with Kepler and could have availed himself of the new insights. In some respects, therefore, when it came to astronomy, he was a little old-fashioned. This despite the fact that it was Galileo who did the third major thing needed to get the Copernican world vision up and running. It was he who provided arguments to show that, counterintuitive though it may be to think that the earth is moving, it really does not violate common sense all that much.

Galileo did this in his famous *Dialogue*, where he provided homely types of examples and analogies, dispelling worries and illusions. Take the concern that if the earth is moving, then objects that are dropped from heights might be expected to lag behind and fall way back from where they are released. He pointed out that on board a ship objects behave as though the ship is stationary even though it may be moving, since it is all a relative rather than an absolute matter. If on this earth objects are moving along with the earth, which is just what the Copernican position does presume, then they are not going to lag behind when dropped. Even more dramatic was the fact that it was Galileo who learned of the telescope and used it to study the heavens. He found such things as the mountains on the moon, thus suggesting that everything up there is not as neat and tidy as presumed by the Aristotelian; that the planet Venus waxes and wanes and changes in apparent size, as would be expected with the heliocentric system but not with the geocentric system; and that Jupiter has moons of its own, proving, among other things, that the earth is not so very special in having a moon and that certainly not everything has the earth as the center of its motion. All of these things were starting to bring the Copernican system into the realm of accepted truth rather than the fantasies of a few odd Platonists.

As is well known, it was doing this work that got Galileo into trouble with the religious authorities, leading eventually to his arrest and his forced recantation of the heliocentric system. It might be thought that this must be a major part of our story, concerned as we are with the relationship between science and religion, and in a sense it obviously is. Later we shall have to study this putative conflict in its own right. But even now we should note that in some ways it was a little bit tangential. The Catholic Church that condemned Galileo was then in the midst of the Counter-Reformation, a time when the church was firmly shutting the stable door after the Protestant horse had fled. It was tamping down on dissent, and this included science, acting now in a way quite unlike the way in which it had supported and cherished science in the Middle Ages. Also, Galileo himself did not help matters, because not only was he now publishing in the vernacular (Italian) so that the ideas could be more widely disseminated, he was also sneering at the Aristotelian authorities, especially in his *Dialogue*, portraying them as naïve and stupid. None of this is to excuse what happened to him, but it is to start to put things in context and plead that one should look carefully at claims that the Galileo affair shows that science and religion will always necessarily be in opposition.

DESCARTES AND NEWTON

Building on the work that had gone before, the seventeenth century then saw two major attempts to make sense of everything and to give a causal understanding of the Copernican move to demote the status of the earth and to give it to the sun. First there was the great French philosopher, mathematician, and scientist René Descartes (Garber 1992). By the seventeenth century there had been a strong revival of Greek atomism, which saw all matter as made of particles moving around in the void. For reasons to be discussed shortly, Descartes was with Aristotle in wanting to deny the void, but he took and modified atomism to his own ends in a form of corpuscular theory. He believed that the whole of space is filled with matter of some kind, that this matter is infinitely divisible, and hence that all motion is a function of force or pressure being applied to corpuscles at one point in space and then being transmitted directly to corpuscles at another point in space. Building on this belief, he then proposed his vortex theory of the universe. Think of a whirlpool, with boats being dragged around in circles. The boats are moved because they are in contact with the water, and each bit of water moves because it is in contact with other bits of water on all sides. Everything is touching something, and that is why everything keeps moving. A solar system such as ours is like a whirlpool (it is a vortex), with the sun at the center – hot and luminous because that is where the major friction occurs – and the planets like the earth being carried along in the matter that fills the universe. For good measure, the occasional comet that we see in our universe is carried along by the swirling waters until it is flung out into another whirlpool-like universe [Figure 2.2].

This theory was incredibly influential, right through to the middle of the eighteenth century, but then it gave way to a rival theory, proposed towards the end of the seventeenth century by the English scientist Isaac Newton (Westfall 1980). Although there were overlaps in their thinking – Newton, like Descartes, believed in a world of small particles – Newton's approach was very different from Descartes', for he postulated a force possessed by all bodies, leading them to attract all other bodies. This was the basis of Newton's well-known law of gravitational attraction, postulating that the strength of the attraction between bodies is proportional to their size

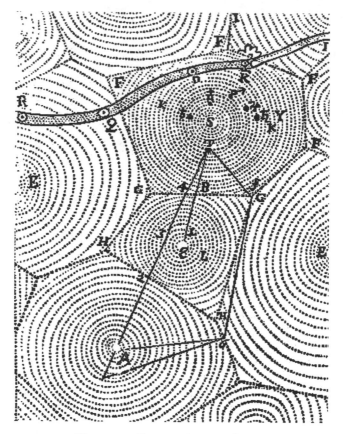

Figure 2.2. Descartes' vortex theory.

(actually, their mass, factoring in density as well as raw linear dimensions) and inversely proportional to the square of the distance between them. In fact, this idea of a gravitational attraction – differing from Descartes in supposing that objects do not have to touch each other to affect each other, allowing for "action at a distance" – was not new to Newton. People had been studying what seemed to be similar forces, notably magnetism, whereby objects attract and repel each other across distances. This had led some – Kepler, for one – to suppose that magnetism or something like it (or in combination with it) might be an important causal factor in keeping the universe in motion and reasonably stable. After all, if one (fairly reasonably) disagreed with Descartes that the universe is filled with some kind of invisible, intangible matter that keeps everything in motion, then forces acting over a distance were just about the only other option. Newton's genius was to take the idea and embed it in a formal system – an empirical axiom system – whereby he could show that with the gravitational law, together with basic laws of kinematics (refined from the kind of work that Galileo and his successors had produced), he could infer the motions of the whole universe. Specifically, he could collapse entirely the Aristotelian distinction between the earth below and the heavens above, showing that one system could simultaneously explain the planetary motions as traced out by Kepler and the motions of terrestrial objects as traced out by Galileo. Finally, the Copernican insights were put on a firm causal basis.

> Newton's work was not perfect, nor was it complete; . . . In the ground he had surveyed he left many blank areas for his successors to fill up, and indeed a number of important mistakes for them to correct. Nevertheless, with his work the scientific revolution reached its climax, and a model for future natural philosophers had been created. Galileo's and Kepler's confidence in the mathematical structure of nature was by it fully justified and 'mechanical principles' were proved to be a sufficient basis for explanation universally throughout physical science. Thus the unity of nature was made manifest in a grand synthesis revealing the applicability of the same laws, the same principles of explanation, in the heavens and on the Earth. The planetary revolutions of Copernicus, Kepler's laws, the discoveries made by Galileo and Huygens relating to the phenomena of gravity and motion, were all shown to follow from these laws and principles, and to be embraced within the same synthesis. (Hall 1983, 306)

THE MACHINE METAPHOR

With hindsight, we can easily see how science had made the move from the organic metaphor to the machine metaphor. We have a switch in root metaphors. We have left the self-regenerating world, where everything is referred to ends, and we are now in a world of particles – atoms, corpuscles – endlessly moving in space – mindlessly, as one might say. Were people back then conscious of this shift? Very much so! One of the more articulate writers on the topic was Robert Boyle, he whose name is attached to one of the most famous laws in the whole of science and a man whose contribution to the Scientific Revolution in Britain was second only to that of

Newton. Boyle penned several essays on what we today would call "philosophy of science," including, most importantly, *A Free Inquiry into the Vulgarly Received Notion of Nature*, a work written mainly in the mid-1660s although not published until 1687. It is in this work that Boyle most openly and in greatest detail announces that the metaphor to substitute for that of Aristotle is that of a machine, a clock or some like artifact. An artifact made by God. Writing against the Aristotelian notion that somehow "nature" itself has a being and a kind of mind or life force of its own, and making reference to a wonderful clock built (between 1571 and 1574) by the Swiss mathematician Cunradus Dasypodius, Boyle responds:

> And those things which the school philosophers ascribe to the agency of nature interposing according to emergencies, I ascribe to the wisdom of God in the first fabric of the universe; which he so admirably contrived that, if he but continue his ordinary and general concourse, there will be no necessity of extraordinary interpositions, which may reduce him to seem as if it were to play after-games – all those exigencies, upon whose account philosophers and physicians seem to have devised what they call nature, being foreseen and provided for in the first fabric of the world; so that mere matter, so ordered, shall in such and such conjunctures of circumstances, do all that philosophers ascribe on such occasions to their omniscient nature, without any knowledge of what it does, or acting otherwise than according to the catholic laws of motion. And methinks the different between their opinion of God's agency in the world, and that which I would propose, may be somewhat adumbrated by saying that they seem to imagine the world to be after the nature of a puppet, whose contrivance indeed may be very artificial, but yet is such that almost every particular motion the artificer is fain (by drawing sometimes one wire or string, sometimes another) to guide, and oftentimes overrule, the actions of the engine, whereas, according to us, it is like a rare clock, such as may be that at Strasbourg, where all things are so skillfully contrived that the engine being once set a-moving, all things proceed according to the artificer's first design, and the motions of the little statues that as such hours perform these or those motions do not require (like those of puppets) the peculiar interposing of the artificer or any intelligent agent employed by him, but perform their functions on particular occasions by virtue of the general and primitive contrivance of the whole engine. (Boyle 1996, 12–13)

Strong stuff! Yet, with respect to the machine metaphor, for our purposes Descartes is an even more interesting and illuminating thinker than Boyle. Well known is the Frenchman's claim that there are two basic kinds of substance, thinking substance (*res cogitans*) and material substance (*res extensa*). They have the key distinguishing attributes of thought and extension. God is pure thinking substance; the physical world is pure material substance; we humans (and perhaps angels) are uniquely in the middle, both thinking and material substance. "Thought and extension can be thought of as constituting the natures of intelligence and corporeal substance, and then they must not be conceived of otherwise than as thinking substance itself and extended substance itself, that is, as mind and body; in this way they will be understood most clearly and distinctly" (Descartes, *Principles*, 1, 63, quoted in Garber 1992, 66). With this kind of philosophy – note, incidentally, that for Descartes a void is logically impossible, because if there is extension there must be substance – the experienced world, other than the part taken up by human minds, simply has to be a blind, unthinking machine. It cannot be anything else.

All of this comes out very clearly in a famous discussion in Descartes' *Discourse on Method*. He discusses the heart and naturally is much taken with the demonstration of the anatomist Harvey that the heart is pumplike (a machine). He then goes on to mention that there are quite a few other things about the human body that seem to happen without being intended, purely automatically.

> This will hardly seem strange to those who know how many motions can be produced in automata or machines which can be made by human industry, although these automata employ very few wheels and other parts in comparison with the large number of bones, muscles, nerves, arteries, veins, and all the other component parts of each animal. Such persons will therefore think of this body as a machine created by the hand of God, and in consequence incomparably better designed and with more admirable movements than any machine that can be invented by man. (Descartes 1964, 41)

Descartes then at once draws the notorious consequence – something that has caused more ill will between the British and the French than even Napoleon – that animals cannot be thinkers and therefore cannot feel pain. "Here I paused to show that if there were any machines which had the organs and appearance of a monkey or of some other unreasoning animal, we would have no way of telling that it was not of the same nature as these animals" (ibid.). Given his philosophy, there was no other conclusion that he could have drawn.

FINAL CAUSES?

Surely, at this point one might want to pause and point out what Descartes admits in the passage just quoted, that because you are adopting a machine metaphor, you are hardly thereby cutting out ends. Now you may not be concerned with the animal or vegetable ends of the world, but you are concerned with God's ends. The world is God's machine, God's artifact, and as such has a creator and a purpose. We make a machine for our ends – the Strasbourg clock to tell the time – and so God makes the world for His ends. Most obviously, God made the world as a place for His favored creations, the beings made in His image, humans, and consequently (even if God had not wanted it) the human drama of Fall and Redemption. The seventeenth-century response was that all of this is true. However, for the working scientist it might be that searching for immediate ends, final causes in the Aristotelian sense, would not be very helpful. This is especially so if the world is no longer considered an organism. Better at first to concentrate on how things are working rather than on why they are working. This seems to have been the position of the great philosopher of the Scientific Revolution, the English politician Francis Bacon in his *Advancement of Learning* (1605). He likened final causes to Vestal Virgins, decorative but sterile!

Boyle had a rather different argument, namely, that it was best to keep God out of the scientific equation altogether. Otherwise, too often you will put things down to miracle, when more hard work would show that law is operating.

> And when I consider how many things that seem anomalies to us do frequently enough happen in the world, I think it is more consonant to the respect we owe to divine providence to conceive that, as God is a most free as well as a most wise agent, and may in many things have ends unknown to us, he very well foresaw and thought it fit that such seeming anomalies should come to pass, since he made them (as is evident in the eclipses of the sun and moon) the genuine consequences of the order he was pleased to settle in the world, by whose laws the grand agents in the universe were empowered and determined to act according to the respective natures he had given them; and the course of things was allowed to run on, though that would infer the happening of seeming anomalies and things really

repugnant to the good or welfare of divers particular portions of the universe. (Boyle 1996, 13)

Descartes, although he was not always quite as free from final-cause thinking as he assumed, argued that it is better to stay away from final causes because we can never be completely sure of God's intentions.

> And so . . . we should not take any reasons for natural things from the ends which God or nature proposed for themselves in making them, since we should not glorify ourselves to such an extent that we think we are privy to their counsel. But considering him [God] as the efficient cause of all things, we shall see what it appears we must conclude by the light of reason he gave us, from those of his attributes he wanted us to have some notion of, about those effects of his which appear to our senses. (*Principles* 1, 28, quoted in Garber 1992, 273–4)

We start to see how God gets pushed out of the equation. Not that Descartes (or Boyle, for that matter) had any intention of threatening the actions or the all-importance of God. As Descartes wrote about God to Princess Elisabeth of Palatine in 1845, "one can only prove that he exists by considering him as a supremely perfect being, and he would not be supremely perfect if something could happen in the world that did not derive entirely from him. . . . God is the universal cause of everything in such a way that he is the same way the total cause of everything and thus nothing can happen without his will" (Garber 1992, 300). However, it is the thin end of the wedge. One is hardly surprised to hear that Newton, having denied the Trinity (on biblical rather than scientific grounds and, prudently, very much in private), rather thought of God in the way of the deists, as an unmoved mover who sets things in motion and then stands back and lets things happen. As it happens, Newton was not completely consistent here and did think that every now and then God had to fiddle with his creation to get things right. But the point is that God was now out of the day-to-day workings of the world. More on this point in a moment.

RECOGNIZING METAPHOR

The main thought I am trying to draw here, the conclusion I want us to take with us, is not that any of this is good, bad, or indifferent. I am not at this point bothered about what did or did not happen to

God. The point I want to make is that we have a metaphor here – the metaphor of the world as a machine. The world is not really a machine, meaning a human-made artifact. It is not a clock or a sewing machine or an automobile. But starting with the Scientific Revolution, we treat it as one. And notice that this has interesting consequences. There are going to be limits, either forced on us by the metaphor or coming from us owing to our own interests: limits that dictate what the metaphor can or is allowed to do. If I say that my love is a rose, then I am not saying anything about her mathematical abilities. She might be a wiz at math or she might be lousy. The metaphor tells me nothing. If I say my love is a rose, then it is highly unlikely that I am interested in whether she is infected by parasites as a rose might be infected by aphids. I could explore that, but I doubt that I will – not if I want her to stay my love! And then of course there are some things on the border. I doubt that when I say my love is a rose I am saying she is a bit on the prickly side. Although I could possibly be, if I am joking. (Joking or not, I suspect that in this case I am trying to say something I think true.)

The point I am making is that we have a metaphor, and metaphors are not absolute. On their own they do not dictate truth. Truth and insight come from the ways in which we use (or are used by) metaphor. This means that you need to be very careful about what you say on the topic of God on the basis of the machine metaphor. Nothing follows simply. Certainly nothing follows as simply as all of those very confident scientists in the Introduction would have it. What might be happening to God, in the sense of affirming or denying His existence, might not be because of God but because of the metaphor – which you might or might not be using in ways that others find compelling. This does not let God entirely off the hook. It does not mean that nothing in science is relevant to the God question, His existence and His nature. If a metaphor works, there is probably a good reason. But we should be careful of assuming that we have absolute logical proof.

I shall be returning to this in much more detail later. For now, I want to make one final reinforcing historical point. The metaphor of the world as a machine had to be stretched somewhat in the seventeenth century, extending what one would mean by "machine," to bring in Newton's gravitational force, action at a distance. For Descartes and his followers, it did not matter how good a theory like Newton's might be, it could not qualify as a genuine scientific theory because it was not sufficiently machinelike. It appealed to occult forces. In his *Rules for the Direction of the Mind*, written fifty years before Newton published his theory (in his *Principia*), Descartes was explicit on this. "Thus, for example, if I should wish to inquire whether any natural force can travel instantaneously to a distant point and pass through all intermediate points, I will not immediately consider magnetism or the influence of the stars, nor even the velocity of light . . . but I will rather think about the motion of bodies in space, because nothing in this category can be more apparent" (Descartes 1964, Rule IX, 181). Likewise, we might want to think about a long rod – when one end moves, so does the other in tandem. It is acceptable to say that moving one end moved the other end, because there are connections all down the line. No one is appealing to forces bridging the gap between beginning and end.

> In the same way, if I wish to learn how contrary effects can be produced simultaneously by one and the same simple cause, I will not borrow examples from the doctors, whose drugs expel certain humors and retain others; nor will I talk nonsense about the moon, saying that it heats by its light and cools by an occult quality; but rather I will consider a balance, in which the same weight causes one pan to rise at one and the same instant that it depresses the other, and other similar cases. (Ibid.)

Newton was not insensitive to this kind of critical attack, especially since there was probably some truth in it! We know that he took an intense interest in alchemy, and it is at least possible that the occult forces he would have encountered in that area of inquiry seeped over into his more respectable work in physics. Be that as it may, he had to make some answer. And what could he do but reply that essentially he had no idea about the nature of gravity but that he was nevertheless allowed to use it?!

> Hitherto we have explained the phenomena of the heavens and of our sea by the power of gravity, but have not yet assigned the cause of this power. This is certain, that it must proceed from a cause that penetrates to the very centers of the sun and planets, without suffering the least diminution of its force. . . . But hitherto I have not yet been able to discover the cause of those properties of gravity from phenomena, and I frame no hypotheses [*hypotheses non fingo*]; for whatever is not deduced from the phenomena is to be called an hypothesis; and hypotheses, whether metaphysical or physical, whether of occult properties or mechanical, have

no place in experimental philosophy. In this philosophy particular propositions are inferred from the phenomena, and afterwards rendered general by induction. Thus it was that the impenetrability, the mobility, and the impulsive force of bodies, and the laws of motion and of gravitation, were discovered. And to us it is enough that gravity does really exist, and act according to the laws which we have explained, and abundantly serves to account for all the motions of the celestial bodies, and of our sea. (From the *Scholium Generale* concluding the *Principia*, quoted in Dijksterhuis 1961, 481)

Actually, Newton did go straight on to frame hypotheses, speculating "about a most subtle spirit which pervades and lies in all gross bodies; by the force and action of which the particles of bodies attract one another at near distances, and cohere, if contiguous; and electric bodies operate to greater distances, as well as repelling and attracting the neighboring corpuscles; and light is emitted, reflected, inflected, and heats bodies; and all sensation is excited, and the members of animal bodies move at the command of the will" (p. 484). Ultimately, however, what sold people on Newton's theory was the fact that it worked – it unified, and it allowed predictions of an accuracy hitherto undreamed of. Descartes may have given a theory more intuitively satisfying, but Newton gave a theory that delivered the goods. People had to put up with gravity, and sooner or later they did.

And this point confirms what has been said about metaphor in this chapter. The "world as a machine" is not something discovered, something given to us by nature. It is something we create in conjunction with nature. How far it applies, what exactly it implies and answers, what it rules out of discourse, how it is to be stretched or not, are matters for negotiation. We see this most clearly with respect to final causes. Man-made machines are teleological. Clocks are for telling the time. Automobiles are for taking people from one place to another. If the physical world is a machine, then what is it for? What are the parts for? Springs in clocks are for saving energy. Wheels on automobiles are for moving rapidly over the ground. What is the sun for? What is the Mississippi for? The scientists of the Revolution found that these sorts of questions did not yield answers – at least they did not yield answers they could use in their science. Increasingly they wanted to keep God at a distance. It was nothing personal. It was that miracles and the like got in the way. This was not something absolute or provable a priori. Newton thought that God did intervene and correct things. But as the science succeeded and the need for miracles diminished, the willingness even to contemplate miracles became increasingly small. Hence, the machine metaphor had to be revised, or restricted. It was fine to ask how things work. What does the spring do? What do the wheels do? It was not fine to ask what ends the working things serve.

Hence in physics – and note the qualification – the machine metaphor, let us now call it the "mechanist" metaphor, was strictly nonteleological. It was a machine metaphor, but it was a restricted metaphor: "the science called mechanics had emancipated itself in the seventeenth century from its origin in the study of machines, and had developed into an independent branch of mathematical physics dealing with the motion of material objects and finding in the theory of machines only one of its numerous practical applications" (Dijksterhuis 1961, 498). Ultimately the justification was pragmatic. If it works – and it does – then good. And, for the scientists, that was enough. But in our inquiry we are not scientists. We are people looking at science. Hence, because the metaphor works, that does not mean that we can or should forget the history, and there are times when it might prove appropriate to go back to the origins and see what routes were taken and what routes not taken. When we do that, we may see that what habit and success leads us to think inevitable, or (far worse) "reasonable," isn't nearly as inevitable or reasonable as we assume. There may be other positions that we could take that are not as silly as we often think. Or there may be implications from the pragmatic choices that are more significant than we generally are prepared to allow.

THREE

ORGANISMS AS MACHINES

We are survival machines, but 'we' does not mean just people. It embraces all animals, plants, bacteria, and viruses. The total number of survival machines on earth is very difficult to count and even the total number of species is unknown. Taking just insects alone, the number of living species has been estimated at around three million, and the number of living insects may be a million, million, million.

Different sorts of survival machines appear very varied on the outside and in their internal organs. An octopus is nothing like a mouse, and both are quite different from an oak tree. Yet in their fundamental chemistry they are rather uniform, and, in particular, the replicators which they bear, the genes, are basically the same kind of molecule in all of us – from bacteria to elephants. We are all survival machines for the same kind of replicator – molecules called DNA – but there are many different ways of making a living in the world, and the replicators have built a vast range of machines to exploit them. A monkey is a machine which preserves genes up trees, a fish is a machine which preserves genes in the water; there is even a small worm which preserves genes in German beer mats. DNA works in mysterious ways.

Dawkins 1976, 22

If Richard Dawkins ever turned to philosophy for enlightenment, one suspects that he would be strongly attracted to the thinking of René Descartes. We have seen already the great French philosopher's thinking about organisms and machines. This was no aberration. Here, from his *Treatise of Man*, is a little thought experiment in which Descartes invites us to join:

> I make the supposition that the body is nothing else but a statue or earthen machine, that God has willed to form entire, in order to make it as similar to us as is possible. Thus he not only would have given it the external color and shape of our members, but also he put in the interior all the parts which are required to make it walk, eat, respire, and that it imitate, in the end, all of our functions which can be imagined to proceed from matter alone, and depend only on the disposition of the organs. (Descartes 1664, 1–2)

Descartes then runs through all of the workings of this machine, concluding:

> I would like you to reflect, after the preceding, on how all the functions that I have attributed to this machine, such as the digestion of food, the beating of the heart and arteries, the nutrition and growth of the members, respiration, waking and sleep, reception of light, sound, smell, taste, heat and such qualities by the external sense organs, the impression of these sensory ideas on the organ of common sense and imagination [i.e., the pineal gland], and the retention or imprinting of these ideas in the memory, occur. Similarly, reflect on the internal motions of the appetites and passions, and finally on the external motions of all the members, which follow with reference both to the objects presented to the senses, and to the passions and impressions contained in the memory, which are imitated as closely as possible those of a true man. Thus, I say, when you reflect on how these functions follow completely naturally in this machine solely from the disposition of the organs, no more nor less than those of a clock or other automaton from its counterweights and wheels, then it is not necessary to conceive on this account any other vegetative soul, nor sensitive one, nor any other principle of motion and life, than its blood and animal spirits, agitated by the heat of the continually burning fire in the heart, and which is of the same nature as those fires found in inanimate bodies. (p. 106–7)

SENSES OF MECHANISM

Actually, however, the story is more complex than you might think. There is no straight line from René Descartes writing in the early part of the seventeenth century to Richard Dawkins writing in the later part of the twentieth century. Robert Boyle was already signaling this before the seventeenth century was out. Descartes wanted to expel final causes from science because we can never know their nature. Although he was onside with the inorganic world, when it comes to organisms Boyle thought it ridiculous to pretend that there are no final causes – or that we cannot ever know the nature of these final causes, that is, of God's intentions. If the eye is not made for seeing, then absolutely nothing makes sense at all. Against the Frenchman and his followers ("Cartesians"), it is a positive moral obligation to study nature and to work out its adaptations. As Boyle wrote in another of his essays, the "Disquisition about the Final Causes of Natural Things":

> For there are some things in nature so curiously contrived, and so exquisitely fitted for certain operations and uses, that it seems little less than blindness in him, that acknowledges, with the Cartesians, a most wise Author of things, not to conclude, that, though they may have been designed for other (and perhaps higher) uses, yet they were designed for this use. As he, that sees the admirable fabric of the coats, humours, and muscles of the eyes, and how excellently all the parts are adapted to the making up of an organ of vision, can scarce forbear to believe, that the Author of nature intended it should serve the animal to which it belongs, to see with.

Boyle continued that supposing that "a man's eyes were made by chance, argues, that they need have no relation to a designing agent; and the use, that a man makes of them, may be either casual too, or at least may be an effect of his knowledge, not of nature's." But not only does this then take us away from the urge to dissect and to understand – how the eye "is as exquisitely fitted to be an organ of sight, as the best artificer in the world could have framed a little engine, purposely and mainly designed for the use of seeing" – it also takes us away from the designing intelligence behind it (Boyle 1688, 397–8).

Note, the important thing is that Boyle did not see this position of his as something threatening to the mechanical position but as complementing it. Moreover, one should see that he distinguished between acknowledging the use of final causes *qua* science, and the inference *qua* theology from final causes to a designing God. First: "In the bodies of animals it is oftentimes allowable for a naturalist, from the manifest and apposite uses of the parts, to collect some of the particular ends, to which nature destined them. And in some cases we may, from the known natures, as well as from the structure, of the parts, ground probable conjectures (both affirmative and negative) about the particular offices of the parts" (p. 424). Then, the science finished, one can switch to theology: "It is rational, from the manifest fitness of some things to cosmical or animal ends or uses, to infer, that they were framed or ordained in reference thereunto by an intelligent and designing agent" (p. 428). To complexity (what Boyle would call "contrivance") in the realm of science, and then to design in the realm of theology.

In the study of organisms, final causes – a legacy (some might say hangover) from the old organic metaphor – were still around. The machine metaphor could not move right in. And this is the topic of this chapter. Did the machine metaphor eventually succeed? And what exactly would it mean to say that it did succeed? Drawing on what we have learned from the last two chapters, there are three possibilities. First, there is what we might call the original position, the position before the Revolution. Organisms are living things, and cannot be reduced to blind, unthinking matter. Remember that Aristotle thought that they had "souls," not in the Christian sense – we shall be discussing this sense in detail in later chapters – but in the sense of some kind of life force. This is what we might call a "vital" force, and hence the position generally can be called "vitalism." Then, at the other extreme, we have the position that Descartes perhaps wanted – if we interpret his talk about functioning not in terms of ends but simply in terms of working (as in, "the spellcheck in this word-processing program really doesn't function that well") – that is, the hard-line mechanistic position. Organisms are in this respect identical to (and require no more understanding than) inorganic objects like planets. We can think of them mechanistically, and there is no need for final causes. Then perhaps in the middle we have a machine-metaphor position, but one that nevertheless wants to talk in terms of ends, of purposes. One that respects and uses final causes in some sense. One that continues to use the full-blooded metaphor of a machine, an artifact, as in: "this keyboard is for typing."

In distinguishing these two senses of the machine metaphor, I am doing nothing new. The *Concise Oxford Dictionary* is ahead of me: "**mechanism:** 1 the structure or adaptation of parts of a machine. 2 the mode of operation of a process." (Mechanism 1 is my in-between option; mechanism 2 is what, in the last chapter, I have simply called "mechanism" without qualification.) The philosophers of science have also beaten me to it. Rom Harré warns that caution is needed "as to the meaning of 'mechanism.'" He continues:

> In ordinary English this word has two distinct meanings. Sometimes it means mechanical contrivance, a device that works with rigid connections, like levers, the intermeshing teeth of gears, axles, and strings. Sometimes it means something much more general, namely any kind of connection through which causes are effective. . . . It is this latter sense that the word is used in science generally, in such diverse expressions as the mechanism of the distribution of seeds and the mechanism of star formation. In hardly any of these cases is any mechanical contrivance being referred to. So we must firmly grasp the idea that not all mechanisms are mechanical. (Harré 1972, 118)

This last passage rather presupposes what we must find out. Clearly Boyle would disagree if "used in science generally" is taken to mean "preferentially" or even "almost invariably." We have to see if there was a move from vitalism to some form of machine metaphor, and if there were reasons why the move was not all the way to full-blooded mechanism (mechanism 2 in the dictionary sense – what we can continue to call "mechanism" without qualification). And if there was not such a move, stopping rather at machines with final causes (mechanism 1 – and what we can call "artifact mechanism"), the objections of the physicists of the seventeenth century could be answered. Can one do predictive, rigorous science? Does the final cause talk do real work, or is it just metaphysical froth on the quality science below? Is a midway position both necessary and viable?

FROM MECHANISM TO VITALISM

In 1727, the Armenian-born, Italian-residing medical researcher Georgio Baglivi made a defiant statement of the mechanistic view of life:

> Whoever examines the bodily organism with attention will certainly not fail to discern pincers in the jaws and teeth; a container in the stomach; watermains in the veins, the arteries and the other ducts; a piston in the heart; sieves or filters in the bowels; in the lungs, bellows; in the muscles, the force of the lever; in the corner of the eye, a pulley, and so on. So let the chemists continue to explain

natural phenomena in complex terms such as fusion, sublimation, precipitation, etc., thus founding a separate philosophy. It remains unquestionable that all these phenomena must be seen in the forces of the wedge, of equilibrium, of the lever, of the spring, and of all the other principles of mechanics. In short, the natural functions of the living body can be explained in no other way so clearly and easily as by means of the experimental and mathematical principles with which nature herself speaks. (G. Baglivi, *De Praxi medica*, in *Opera omnia medico-practica et anatomica* [Venice, 1727], quoted in Moravia 1978, 48)

This is perhaps more an expression of artifact mechanism than of full-blooded mechanism, but we do seem to be well on the way to the tougher position found in the writings of the atheistic French physician Julien Offray de la Mettrie and his *L'Homme Machine* (1748). "The human body is a machine which winds its own springs. It is the living image of perpetual movement. Nourishment keeps up the movement which fever excites. Without food, the soul pines away, goes mad, and dies exhausted" (La Mettrie 1912, 93). Not that national differences have no causal role to play. A good meal "infects the souls of comrades, who express their delight in the friendly songs in which the Frenchman excels." Across the Channel, however, we find "the English who eat meat red and bloody" and who "seem to share more or less in the savagery due to this kind of food" (p. 94). Stirring stuff, yet already before the beginning of the eighteenth century the empirical researchers were starting to find the purely mechanistic view – even an artifact mechanistic view – inadequate. In 1685, the witty French essayist Bernhard de Fontenelle, in his *Lettres gallants du Chevalier d'Her*, put his finger on one of the major problems, the matter of reproduction: "You say that beasts are machines, just like watches? Yet if you put a male dog machine next to a female dog machine, a third little machine could result from them; two watches, however, could rest next to each other for their entire existence without ever making a third watch." Concluding: "I find through our philosophy that all things that are two but have the power to make three are of a quality elevated well above that of a machine" (quoted in Roget 1997, 118).

One who stood strongly against the mechanist view (certainly against mechanism 2, the "mechanism all the way" view) was Georg Ernest Stahl, the German physician and chemist. Although today he is chiefly remembered for his enthusiasm for the now-obsolete phlogiston theory of chemistry, in physiology he was highly influential as one insisting on special "animist" or (to think in traditional terms) "vital" forces peculiar to living things. These are forces that give order or purpose, in other words, that speak to Aristotle's final causes. "Before anything else, we must know the exact meaning of what is vulgarly called life. What does it formally consist of? What is its function, both from the material and subjective point of view, and from the final and objective point of view? What use has it, what need is there for it in the body? What does it offer? And how useful or even necessary is it?" (*Theoria medica Vera* 14, quoted in Moravia 1978, 50)

Stahl had great influence down in Montpellier, the home of one of France's oldest and greatest medical schools. So too did the German anatomist Albrecht von Haller, just then arguing that there are forces or notions of irritability (causing parts of the body to contract when touched) and sensitivity (transmitting sensations through nerves to the brain) that are peculiar to living things. This was the position of the physician and friend of the Encyclopedist Denis Diderot, Theophile Bordeu: "Each living part of the body has nerves which give a sensibility, a species or particular degree of sentiment." He gave a vivid metaphor for how he saw life being organized in a way very unlike that of machine action.

> We compare the living being, in order better to understand the particular actions of each part, to a swarm of bees, which gather themselves into a ball and which suspend themselves from a tree in the manner of a bunch of grapes.... So, in order to follow the comparison of the bunched up bees, it is completely glued to the branch of the tree by the action of the bees which must work together in order for it to hang on; everything works together to form a solid body and each part depends on the actions of the other parts. The analogy is obvious; the parts of the body are next to each other; they each have their own place and their own function; the workings together of these functions, the harmony which results, gives health. If the harmony is disturbed, thanks to one part not working, which puts other parts out of order or functioning, so that the actions are backed and things cannot work properly, then these changes bring on bad health of greater or lesser intensity. (Moravia 1978, 56, quoting T. Bordeu, *Recherche anatomiques sure la position des glandes*, 1, 187, in *Oeuvres complete* [Paris, 1828])

Organisms are not machines. They are sensitive beings (*êtres sensibles*).

This kind of thinking tied in with ideas about individual development. In the seventeenth and eighteenth centuries there were bitter debates about the nature of fertilization and the origins of new organisms. On the one side were the "preformationists," who argued that life forms come from previous life forms and that, rather like one of those Russian dolls, the fully formed organism can be found in the sex units (some opted for sperm, some for ova), and within that can be found another fully formed organism, and so on ad infinitum. (For someone like Descartes, who thought that matter is infinitely divisible, this is no problem. It is obviously more problematic for the corpuscular theorist, who had particles moving in the void.) On the other side were the "epigeneticists," who thought that life forms were created anew in each generation. This was not a claim that life itself was created anew – that had to come from previous living material – but that that what Aristotle thought was covered by formal causes, the shape and nature, had to come anew.

In the first flush of the success of the Scientific Revolution and under the influence of Descartes' philosophy, preformation had the upper hand. It seems clearly to be the method favored by a mechanistic world. There would be no need for occult forces shaping and forming organisms. But by the middle of the eighteenth century, it was clear that preformation had little empirical support, and – most particularly thanks to the work of Caspar Friedrich Wolf – epigenesis came back into favor. In France, the great naturalist Buffon postulated the interior mold (*moule intérieur*), a kind of equivalent to the Aristotelian formal cause. This sets the pattern. Then a special vital force peculiar to the living world, the "penetrating force" – something akin to the Newtonian force of gravity – packs the molecules in according to the mold, or at least the penetrating force orders the molecules in the right way according to the interior mold. In Germany, epigenetic thinking led the physiologist and anthropologist Johann Friedrich Blumenbach to posit the idea of a vital force that forms and shapes the organism, the *Bildungtrieb*.

> First that in all living organisms, a special inborn *Trieb* exists which is active throughout the entire lifespan of the organism, by means of which they receive a determinate shape originally, then maintain it, and when it is destroyed repair it where possible.
> Second that all organized bodies have a *Trieb* which is to be distinguished from the general properties of the body as a whole as well as from the particular forces characteristic of that body. This *Trieb* appears to be the primary cause of all generation,

reproduction, and nutrition. And in order to distinguish it from the other forces of nature, I call it the Bildungstrieb. (Lenoir 1989, 20, quoting Blumenbach 1781)

He admitted that this is a kind of occult force, but (like Buffon) he argued that it is nevertheless also like a Newtonian force.

> The term *Bildungstrieb* just like all other *Lebenskräfte* such as sensibility and irritability explains nothing in itself, rather it is intended to designate a particular force whose constant effect is to be recognized from the phenomena of experience, but whose cause, just like the causes of all other universally recognized natural forces remains for us an occult quality. That does not hinder us in any way whatsoever, however, from attempting to investigate the effects of this force through empirical observations and to bring them under general laws. (p. 21)

KANT ON ORGANISMS

We come now to the work of Immanuel Kant. He was first and foremost a great philosopher, but in the nineteenth century he also had a great influence on the thinking of scientists. Initially convinced of the all-pervading nature of and necessity for Newtonian-type thinking, toward the end of his intellectual career Kant began to think that the life sciences challenged his belief. Could it be that a purely mechanical view of life will not work and that something else, some kind of vital force or forces, was needed for full understanding? Kant was loath indeed to go this far, ontologically at least. But he did start to think that somehow we must allow for something along these lines, as a principle of thinking, if nothing else – as a kind of "pretend" view, adopted to (as we would say) "save the appearances."

The key source is the second part of the *Critique of Judgement* (1790). Agreeing with the biologists of his day, Kant saw that the significant feature of organisms is that they are organized – they and their parts seem as if fashioned for the ends of survival and reproduction. In other words, organisms demand explanation in terms of final causes as well as efficient and other causes. Yet, as a would-be mechanist, Kant saw that final causes are problematic. Apart from anything else, final-cause talk seems to imply conscious design, and this is simply not acceptable in science. We are allowed to talk only in terms of material or mechanical causes. "Hence if we supplement natural science by introducing the conception of God into its context for the purpose of rendering the finality of nature explicable, and if, having done so, we turn round and use this finality for the purpose of proving that there is a God, then both natural science and theology are deprived of all intrinsic substantiality." Kant was unbending on this. "This deceptive crossing and recrossing from one side to the other involves both in uncertainty, because their boundaries are thus allowed to overlap" (Kant 1790, 31).

But Kant recognized that we simply cannot do without final-cause thinking. Heuristically, teleology is absolutely essential in biology. We need the maxim: "*an organized natural product is one in which every part is reciprocally both end and means.*" We simply cannot do biology without assuming final cause. "It is common knowledge that scientists who dissect plants and animals, seeking to investigate their structure and to see into the reasons why and the end for which they are provided with such and such parts, why the parts have such and such a position and interconnexion, and why the internal form is precisely what it is, adopt the above maxim as absolutely necessary." Scientists cannot do biology in any other way. Teleological thinking is not a luxury; it is a necessity. Life scientists "say that nothing in such forms of life is in vain, and they put the maxim on the same footing of validity as the fundamental principle of all natural science, that nothing happens by chance. They are, in fact, quite as unable to free themselves from this teleological principle as from that of general physical science. For just as the abandonment of the latter would leave them without any experience at all, so the abandonment of the former would leave them with no clue to assist their observation of a type of natural things that have once come to be thought under the conception of physical ends" (p. 25).

So how are we to solve the problem, what Kant called an "antimony," of needing to use final-cause talk and yet recognizing that only material-cause talk is acceptable in physical science or in any science that claims to be talking of objective reality? Here the Kantian metaphysics comes into play – phenomenally (experienced reality) we can see no design in nature, but noumenally (absolute reality) it is possible that there is design. God may be standing behind everything, but this is for things in themselves and not for the phenomenal world as we know it.

> It is at least possible to regard the material world as a mere phenomenon, and to think something which is not a phenomenon,

namely a thing-in-itself, as its substrate. And this we may rest upon a corresponding intellectual intuition, albeit it is not the intuition we possess. In this way, a supersensible real ground, although for us unknowable, would be procured for nature, and for the nature of which we ourselves form part. Everything, therefore, which is necessary in this nature as an object of sense we should estimate according to mechanical laws. But the accord and unity of the particular laws and of their resulting subordinate forms, which we must deem contingent in respect of mechanical laws – these things which exist in nature as an object of reason, and, indeed, nature in its entirety as a system, we should also consider in the light of teleological laws. Thus we should estimate nature on two kinds of principles. The mechanical mode of explanation would not be excluded by the teleological as if the two principles contradicted one another. (pp. 65–6)

We may (must) suppose the existence of God, but we cannot prove it. "All that is permissible for us men is the narrow formula: We cannot conceive or render intelligible to ourselves the finality that must be introduced as the basis even of our knowledge of the intrinsic possibility of many natural things, except by representing it, and, in general, the world, as the product of an intelligent cause – in short, of a God" (p. 53).

For Kant, then, teleological thinking is a regulative principle; it is a necessary heuristic. It is not a condition of rational thinking in the way that the mechanical philosophy is. We cannot think of the world other than causally, for instance. We can certainly look at organisms without thinking of final causes. But as soon as we start to study them, to understand them, final-cause thinking comes into play – has to come into play. Final causes are part of the filter, the lens, through which we study the world. They are our doing: similar to things like causality in that we impute them to the world, but less strong than causality because we can think without them, even though we cannot work without them. They are regulative. "Strictly speaking, we do not observe the *ends* in nature as designed. We only read this conception *into* the facts as a guide to judgement in its reflection upon the products of nature. Hence these ends are not given to us by the Object" (p. 53).

In our terminology, Kant wanted to be a hard-line mechanist (mechanism 2), as in physics, but he felt that the organization of the living world made that impossible, so he inclined to a kind of artifact mechanism (mechanism 1). Most obviously, God was the workman, but that kind of supposition in science is impossible. So Kant is left in a rather uneasy bind, wanting mechanism 2, but settling for something else that required a sort of "as if" thinking. Best of all for Kant would have been if he could have seen his way forward to hard-line mechanism (in the organic world), but this was not possible. Indeed, if anything, thanks to the influence of Blumenbach, we might judge that Kant had one if not both feet in the vitalist camp.

> In all physical explanations of organic formations Herr Hofrat Blumenbach starts from matter already organized. That crude matter should have originally formed itself according to mechanical laws, that life should have sprung from the nature of what is lifeless, that matter should have been able to dispose itself into the form of a self – maintaining purposiveness – this he rightly declares to be contradictory to reason. But at the same time he leaves to natural mechanism, under this to us indispensible principle of an original organization an undeterminable and yet unmistakeable element, in reference to which the faculty of matter is an organized body called a *formative force* in contrast to and yet standing under the higher guidance and direction of that merely mechanical power universally resident in all matter. (p. 29)

Kant himself might well have objected to the title "vitalist." With the scientists, he might have said that the formative force is simply the biological equivalent of Newton's ultimately mysterious force of gravitation. This, explicitly, was Buffon's position and Blumenbach's also. But whatever the right category – putting it in terms of the Greeks, one might say that Kant was offering an Aristotelian solution (vital forces) to a Platonic problem (God the designer) – none of this is very satisfactory. Without unduly and anachronistically applying our current knowledge of science to thinking at the end of the eighteenth century, the problem was that Kant was not dealing with a full deck of cards. He did not have the science to make sense of the situation. Famously, Kant denied that one can have a Newton of the blade of grass. So, rather than picking over the bones of the critical philosophy, let us move forward now to the nineteenth century and see how the biologists picked up the challenges. I turn first to physiology and then to inquiry about origins.

GETTING MECHANISM BACK

For our physiological example, let us choose one basic question and the proffered solutions: *animal heat*. If the machine metaphor is going to work anywhere, then it surely has to work here. Mammals are warm-blooded. They maintain temperatures higher than those of their surroundings. In a way, it is pretty obvious what is going on. How do you keep things warm? By burning stuff like wood; and to do this, you need air. Cut off the supply of the latter, and the fire goes out. Mammals clearly do something similar. They take in fuel – food and drink – and they breathe, thus getting the air. Without the food and drink, without the air, they die. Fully supplied, animals stay warm, and – the equivalent of the ashes from the fire – they expel waste through urinating and defecating, not to mention sweating and breathing out. And just as a fire uses up the air – a lighted candle in a closed jar goes out – so the outgoing out air has been used up. People in confined spaces, using air over and over again, suffocate. These are the basics. The problem is filling in the gaps.

There were those who agreed with Kant that final causes are a crucial part of biological understanding. In a move that was hardly surprising, given Kant's own confusions, they stayed with tradition, arguing that one needs special end-seeking causes, causes that are peculiar to life – that is, vital causes. This does not mean that these people were indifferent to the issues of mechanism; they were even prepared to think of organisms as machines in a sense, but machines of a peculiar kind. Certainly not just clocks. We see this kind of combined (conflicted?) thinking in the early nineteenth-century British physiologist James Frederick Palmer.

> The animal body may be regarded as an intimate piece of machinery, arranged in perfect accordancy with the pre–existing properties of matter, so as to admit of the agency of those properties for the accomplishment of its own ends. . . . Physiology, in short, is a complex science, presenting phenomena which proceed from the combined operation of chemical, physical, and vital agencies, harmoniously intermixed, but within which the vital principle holds a sort of supremacy, directing and modifying all the subordinate agencies to its own definite ends, besides producing effects which are not referable to any other power. (Goodfield 1960, 110)

A more significant scientist – one who shows that being vitalist was not synonymous with being bad or unimportant – was the

Kantian-influenced German chemist Justus von Liebig (1803–1873). One of the most crucial of organic chemists, it was he who discovered the importance of nitrogen for plant growth as well as fundamental facts about human nutrition. (Those of us reared in England know him indirectly through his invention of the bouillon cube, *Oxo*.) He too was prepared to use mechanical metaphors to describe the workings of mammalian body. He likened the way in which food is used to the workings of a furnace.

> It is evident, that the supply of heat lost by cooling is effected by the mutual action of the elements of the food and the inspired oxygen, which combine together. To make use of a familiar, but not on that account less just illustration, the animal body acts, in this respect, as a furnace, which we supply with fuel. It signifies nothing what intermediate forms food may assume, what changes it may undergo in the body, the last change is uniformly the conversion of its carbon into carbonic acid, and of its hydrogen into water; the unassimilated nitrogen of the food, along with the unburned or unoxidized carbon, is expelled in the urine or in the solid excrements. In order to keep up in the furnace a constant temperature, we must vary the supply of fuel according to the external temperature, that is, according to the supply of oxygen.
>
> In the animal body the food is the fuel; with a proper supply of oxygen we obtain the heat given out during its oxidation or combustion. (Goodfield 1960, 123)

Yet he too thought that mechanism is not enough. The Kantian demand must be met. Final causes must be acknowledged and respected.

> Respiration is the falling weight, the coiled spring, which keeps the clockwork in motion.
>
> Whatever the role of magnetic or electrical disturbances in the functioning of the organs, the final cause of all these activities is a process of material exchange which can be expressed as the conversion of components of food into oxygen compounds taking place in a definite period of time. (Liebig 1842, 23, quoted in Lenoir 1989, 163)

For this we need a vital force, or *Lebenskraft*.

> The cause of the *Lebenskraft* and the phenomena of life is neither chemical force, electricity, nor magnetism. It is a force that possesses the most universal characteristics of all causes of motion, form and structural change of matter, and yet it is a different force

because it is associated with manifestations that none of the other forces carry in and of themselves. (p. 164)

Others, however, were not prepared to accept this. They appreciated the problems, but wanted an explanation of animal heat in purely physico-chemical terms that truly spelled success for the Cartesian aspirations. Although the problem of animal heat goes back to the Greeks, it starts to pick up steam toward the end of the eighteenth century, and we can readily see why. If (as seems certain) breathing in air is a crucial part of the process, then at the very least one needs a chemistry that can analyze this and start to make sense of what is happening. The older phlogiston theory, thinking that bodies give up a substance (phlogiston) during burning, and that air serves its burning abilities only so long as it is dephlogisticated, was not going to be adequate. One needed modern chemistry, which sees oxygen as a gas in its own right, and given that the father of this theory was the French chemist Antoine-Laurent Lavoisier, it is not surprising that it was he who made the first moves toward a modern, mechanistic understanding of animal heat.

By the end of the eighteenth century there were four main hypotheses: two that saw the heat being generated by friction brought on by the circulation of the blood (one thought the friction was between blood and walls, and the other between particles in the blood), one (endorsed by Ben Franklin) that put the heat down to fermentation, and one that went straight to combustion of food using the breathed-in air. Knowing that he was dealing with a newly discovered gas, oxygen, and that (as with normal burning) in respiration nitrogen does nothing, while the oxygen gets converted to carbon dioxide, this last was the position of Lavoisier. Along with the physicist Pierre Simon de Laplace, he devised a clever piece of experimental apparatus – an ice calorimeter – to show that the combustion/carbon dioxide equation holds in animals even more exactly than one might expect, and precisely along the lines one finds in normal combustion (Coleman 1971) [Figure 3.1].

Now, of course, the question is how exactly (and where) this combustion occurs. In the 1830s, the German physicist and chemist Heinrich Gustav Magnus showed that it cannot be in the lungs, because although the arterial blood and the venous blood both contain oxygen and carbon dioxide, there is more oxygen in the arterial blood (moving away from the lungs) than in the venous blood

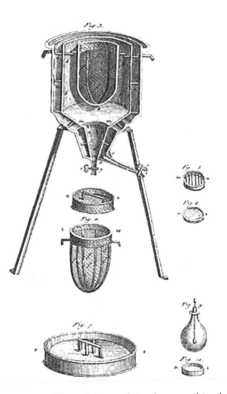

Figure 3.1 Ice calorimeter of Lavoisier and Laplace used in the winter of 1782–3, from the *Elements of Chemistry* (1789) of Lavoisier. (It is in the Conservatoire des Arts et Metiers in Paris.)

(moving toward the lungs). The conversion of fuel into energy must be going on somewhere else in the body. The story now really starts to gather steam, both without and within physiology. Without was the final definitive establishment by the 1840s of the principle of the conservation of energy, a principle that was at once seen (by the practicing physician Julius Robert Mayer, for instance) as directly applicable to organisms: "In the living body, carbon and hydrogen are oxidized and heat and motive power thereby produced. Applied directly to physiology, the mechanical equivalent of heat proves that the oxidative process is the physical condition of the organism's capacity to perform mechanical work and provides as well the numerical relations between [energy] consumption and [physiological] performance" (quoted in Coleman 1971, 123). Within, Liebig was starting to pin things down by showing that although the blood brings along the oxygen, it is probably not the site of combustion. Active muscle tissue uses oxygen and gives up carbon dioxide. This led at once to more mechanical metaphors. In the words of Edward Franklin, a German-trained British chemist:

> A command is sent from the brain to the muscle, the nervous agent determines oxidation. The potential energy becomes actual energy, one portion assuming the form of motion, the other appearing as heat. *Here is the source of animal heat, here the source of muscular power!* Like the piston and cylinder of a steam engine, the muscle is only a machine for the transformation of heat into motion; both are subject to wear and tear, and require renewal; but neither contributes in any important degree, by its own oxidation, to the actual production of the mechanical power which it exerts. (Quoted in Coleman 1971, 134–5)

Obviously, however, things could not stop here. It was at the beginning of the 1840s that cell theory was first properly proposed (by Theodore Schwann and Matthias Jakob Schleiden). The cell is the fundamental unit of life, that is, the fundamental unit from which living things are created and built up. It was then only a matter of time before – especially as in the great work of the French physiologist Claude Bernard – people grasped that combustion is something that occurs within the cell as the molecules of food are broken down and, combining with oxygen (thus releasing carbon dioxide), yield up energy – heat and the other energy needed to supply a functioning, active mammalian body.

Obviously, there is much more to the story than this. For example, when wood burns it becomes very hot, far hotter than the mammalian body could tolerate. Somehow combustion has to take place at much lower temperatures. The key is enzymatic action – some molecules help chemical processes to proceed more rapidly, more often, and at lower temperatures. This was the work of Jons Jacob Bezelius (another Kantian vitalist, incidentally) in the 1830s:

> We may consider the whole of the animal body as an instrument which, from the nourishment it receives, collects materials for continuous chemical processes, and of which the chief object is its own support. But with all the knowledge we possess of the forms of the body considered as an instrument, and of the mixture and mutual bearings of the rudiments to one another, yet the cause of most of the phenomena within the Animal Body lies so deeply hidden from our view, that it will certainly never be found. We call this hidden

cause vital power and like many others, who before us have in vain directed their deluded attention, we make use of a word to which we can fix no idea. (Goodfield 1960, 65)

Likewise there had to be major advances in the understanding of the breakdown of food and of the energies released. This was an area in which Liebig did major work. However, as the century moved toward its close, increasingly physiologists could assert with confidence that the mammal is a heat machine, obeying the same laws of physics and chemistry as anything inorganic. In the last decade of the century, the German physiologist Max Rubner – who, incidentally, was responsible for the claim that smaller animals don't live as long as larger ones because their metabolisms work harder and faster (a claim based on the analogy with machines that work harder and thus wear out faster) – performed a classic series of experiments with a "respiration apparatus," showing that the more or less (and richer or poorer) food an animal received, the more or less heat it gave off. What goes in must come out, and what comes out must have gone in. "Not a single, isolated datum chosen at will out of all these experimental results can leave us in any doubt that the exclusive source of heat in warm-blooded animals is to be sought in the liberation of forces from the energy supply of the nutritive materials" (quoted in Coleman 1971, 142).

There were still vitalists around – in Germany the biologist Hans Driesch, insisting that there is a vital force (an "entelechy"), and in France the philosopher Henri Bergson, postulating "*élans vitaux*." But this sort of stuff was past its sell-by date. In this kind of biology, mechanism had won, especially as micro-studies on the organism gathered speed. Highly instructive is the attitude of Edmund B. Wilson, the American biologist and one of the leaders in the development of cell theory as the nineteenth century turned to the twentieth. In a lecture he gave at Columbia University in 1907, he made clear his enthusiasm for the machine metaphor. He spoke of the "fundamental problem towards which all lines of biological inquiry sooner or later lead us" and then elaborated. "The problem of which I speak is that of organic mechanism and its relation to that of organic adaptation." Continuing: "How in general are the phenomena of life related to those of the non-living world? How far can we profitably employ the hypothesis that the living body is essentially an automaton or machine, a configuration of material particles, which, like an engine or a piece of clockwork, owes its mode of operation to its physical and chemical construction?" He made it clear that the background assumption was not in question. "It is not open to doubt that the living body *is* a machine. It is a complex chemical engine that applies the energy of the food-stuffs to the performance of the work of life" (Wilson 1908, 7–8).

Wilson did not want to leave things there. "But is it [the living body] something more than a machine? If we may imagine the physico-chemical analysis of the body to be carried through to the very end, may we expect to find at last an unknown something that transcends such analysis and is neither a form of physical energy nor anything given in the physical or chemical configuration of the body?" In other words, what about vitalism? "Shall we find anything corresponding to the usual popular conception – which was also along the view of physiologists – that the body is 'animated' by a specific 'vital principle,' or 'vital force,' a dominating 'archæus' that exists only in the realm of organic nature? If such a principle exists, then the mechanistic hypothesis fails and the fundamental problem of biology becomes a problem *sui generis*" (p. 8).

Wilson admitted that he could not answer this question in an absolute way. But for himself, as a practicing scientist, he could see no reason to relinquish the machine metaphor or to restrict its scope. Vital forces do nothing for us as scientists.

> To me it seems not to be science, but either a kind of metaphysics or an act of faith. I must own to complete inability to see how our scientific understanding of the matter is in any way advanced by applying such names as "entelechy" or "psychoid" to the unknown factors of the vital activities. They are words that have been written into certain spaces that are otherwise blank in our record of knowledge, and as far as I can see no more than this. It is my impression that we shall do better as investigators of natural phenomena frankly to admit that they stand for matters that we do not yet understand, and continue our efforts to make them known. And have we any other way of doing this than by observation, experiment, comparison and the resolution of more complex phenomena into simpler components? I say again, with all possible emphasis, that the mechanistic hypothesis or machine-theory of living beings is not fully established, that it *may* not be adequate or even true; yet I can only believe that until every other possibility has reality been exhausted scientific biologists should hold fast to the working program that has created the sciences of biology. The vitalistic

hypothesis may be held, and is held, as a matter of faith; but we cannot call it science without misuse of the word. (pp. 20–1)

The machine metaphor, then – but in what sense was someone like Wilson using the machine metaphor? Was he thinking of organisms as artifacts (mechanism 1) or simply of processes (mechanism 2)? The great twentieth century evolutionist Ernst Mayr (1988) used to distinguish between proximate-cause thinking in biology and ultimate-cause thinking. Proximate-cause thinking is looking at processes (mechanism 2), whereas ultimate-cause thinking is seeking final causes (mechanism 1). Mayr's contention was that the kind of work we have been considering – animal heat, issues to do with the working of the cell, and so forth – is proximate-cause thinking, and there is truth in this. But is it the whole truth? Would it be right to go down the path that a reading of Mayr might seem to indicate, and to suggest that one kind of biology (physiology) uses one kind of explanation (proximate-cause) exclusively and another kind of biology (evolution) uses another kind of explanation (ultimate-cause) exclusively? Let us put off trying to answer this question until we have looked at the biology of origins.

EVOLUTION

Descent with modification. The view that all organisms, living and dead, are the end products of a slow process of change, from other forms, ultimately from the most simple of all and perhaps indeed from inorganic matter. Although there were proto-evolutionary hypotheses even as far back as the Greeks, it was an idea that did not start to emerge in its own right until the eighteenth century. Its appearance was directly a function of the fact that it was then that some people started to embrace a philosophy of Progress – the idea that humans unaided can improve their lot through science and technology and education and the like – and to turn their backs on a philosophy or theology of Providence – the idea that humans are mired in sin and can do nothing but for the redeeming blood of Jesus. Our depravity and His grace. Progress is a view of society that sees people striving and moving upwards; culture or civilization in the broadest sense gets better, and evolution (although it was not called this until the middle of the nineteenth century) piggy-backed on this view, with organisms seen as striving and moving up the chain of being or life. From monad to man, as people used to say (Ruse 1996).

The French biologist Jean Baptiste de Lamarck put things crisply and bluntly:

> Ascend from the simplest to the most complex; leave from the simplest animalcule and go up along the scale to the animal richest in organization and facilities; conserve everywhere the order of relation in the masses; then you will have hold of the true thread that ties together all of nature's productions, you will have a just idea of her *marche*, and you will be convinced that the simplest of her living productions have successively given rise to all the others. (Lamarck 1802, 38)

Along with this naturally tended to come the drawing of analogies between the growth of the individual and the growth of the group. Thus Denis Diderot, the eighteenth-century French *philosophe*: "Just as in the animal and vegetable kingdoms, an individual begins, so to speak, grows, subsists, decays and passes away, could it not be the same with the whole species?" (Diderot 1943, 48).

It is important to stress that the people who became evolutionists were rowing against the tide, fueled in their thinking by strong ideological reasons – reasons that overcame the serious objections to evolution that many serious thinkers could see. Again, Immanuel Kant is seminal. He did not think the idea of evolution was silly, but final cause simply made it impossible. There is no way that blind law can lead to organization and functioning. Organisms are not material machines (Ruse 2006). This kind of thinking was taken up by the great French comparative anatomist Georges Cuvier. (There was almost certainly a direct Kantian influence on this German-educated biologist.)

> Natural history nevertheless has a rational principle that is exclusive to it and which it employs with great advantage on many occasions; it is the *conditions of existence* or, popularly, *final causes*. As nothing may exist which does not include the conditions which made its existence possible, the different parts of each creature must be coordinated in such a way as to make possible the whole organism, not only in itself but in its relationship to those which surround it, and the analysis of these conditions often leads to general laws as well founded as those of calculation or experiment. (Cuvier 1817, 1, 6)

Cuvier gave empirical reasons why evolution cannot occur – he pointed to the mummified animals brought back to France from Egypt by savants following the invading Napoleon, and argued that since they are like today's organisms, even though very old, natural change obviously does not occur. But truly, for Cuvier, evolution is impossible because the move from a functioning organism of one kind to a functioning organism of another kind would demand non-functioning intermediates – literally neither fish nor fowl – and this is impossible. Cuvier was hazy about where organisms ultimately came from – although a practicing Protestant, he did not want to make direct appeals to scripture for such things – but the point was that there was more to the story than just mechanism. Development could not lead to functioning machines.

This was the challenge faced by Charles Darwin. And this was a challenge tackled head on by Charles Darwin, who proposed his mechanism of natural selection not merely to explain organic change, but to explain the nature of organic change – something that pre-serves and promotes adaptation, the features of organisms high-lighted by final-cause thinking, namely, the features that are orga-nized and function to benefit their possessors (Ruse 1979). Thanks to potential population growth outstripping the potential supply of food and space, there is a struggle for existence – or, more perti-nently, a struggle for reproduction. More organisms are born than can survive and reproduce. Success in the struggle, on average, is due to the special features possessed by the winners, or the success-ful (the fittest), and not by the losers, or the unsuccessful (the unfit), and this leads to a natural form of winnowing, or selecting, akin to the practices of animal and plant breeders. There is change, but change of a specific kind, namely, change that produces organisms functioning with features that help – features that seem designed to serve the ends of the organisms, and which to this point seemed to require Aristotelian final causes for their explanation.

So where did Darwin stand on the machine metaphor? Where was he on the various senses of mechanism? As you will appreciate in the light of the brief general discussion about the nature of metaphor in the previous chapter, Darwin never, ever, wanted to deny the designlike nature of organisms. (In a later chapter, we will encounter those who do.) Darwin was quite comfortable using the language of "final cause." But this understanding in no way conflicts with Darwin's aim of using natural selection to give purely naturalistic explanations. He wanted to have nothing to do with occult forces or with God intervening or anything like that (Ruse 2003). This is not to say that Darwin was an atheist. At least through the writing of the *Origin of Species* (1859), in which he announced his theory of evolution through natural selection, he was a kind of deist, thinking of God as an Unmoved Mover. But the processes of evolutionary biology had to be completely natural. Writing in praise of his grand-father, Erasmus Darwin, Charles said: "In contrast to the old theory that all adaptation to purpose in the arrangements of the world was fore-calculated and fore-ordained, and that all organisms were merely wheels in a gigantic machine made once for all, and incapable of improvement, this new view is so grand that it deserved a higher appreciation than it has ever met with. The Cartesio-Paleyan comparison of Nature with a great piece of clockwork (a fundamentally mistaken comparison, because every complete mechanical work has only been attained by many gradual improvements in the course of generations), is finally got rid of by it" (Darwin 1879, 224). In other words, Charles Darwin was totally committed to the mechanistic view of the world, machine metaphor 2, where the idea of final cause or purpose is illicit.

Yet at the same time he was also committed to machine metaphor 1, the artifact metaphor, when it came to thinking about individual organisms. Repeatedly, before he had thought things through in evolutionary terms and after he became a full-blown evolutionist, Darwin thought of organisms in terms of machinelike parts, in terms of the "contrivances" that humans design and build to effect desired ends. It was for him what I have called a root metaphor. Thus in the *Voyage of the Beagle*, in a nonevolutionary context, Darwin describes a luminous beetle or elator, *Pyrophorus luminosus*. He is interested in the springing powers of this creature.

> The elater, when placed on its back and preparing to spring, moved its head and thorax backwards, so that the pectoral spine was drawn out, and rested on the edge of its sheath. The same back-ward movement being continued, the spine, by the full action of the muscles, was bent like a spring; and the insect at this moment rested on the extremity of its head and wing-cases. The effort being suddenly relaxed, the head and thorax flew up, and in consequence, the base of the wing-cases struck the supporting surface with such force, that the insect by the reaction was jerked upwards to the height of one or two inches. The projecting points of the thorax,

and the sheath of the spine, served to steady the whole body during the spring. In the descriptions which I have read, sufficient stress does not appear to have been laid on the elasticity of the spine: so sudden a spring could not be the result of simple muscular contraction, without the aid of some *mechanical* contrivance. (Darwin 1845) [Here, and in subsequent quotations, I have italicized the language of machine or mechanism.]

Notice that, as with Boyle, it is in the final-cause context, the adaptation context, that Darwin uses the language of mechanism. Since natural selection is intended to explain adaptation, we would expect Darwin to use the metaphor in such contexts – and he does. It is artifact mechanism that Darwin is endorsing and working within, not simple mechanism alone. In the *Origin*, for instance, Darwin sets up the problem in this language:

It may be doubted whether sudden and considerable deviations of structure such as we occasionally see in our domestic productions, more especially with plants, are ever permanently propagated in a state of nature. Almost every part of every organic being is so beautifully related to its complex conditions of life that it seems as improbable that any part should have been suddenly produced perfect, as that a complex *machine* should have been invented by man in a perfect state. (Darwin 1872, 34)

And in a little book on orchids, written just after the *Origin* – the most important of all of Darwin's writings for showing how he thinks that selection will actually work – he constantly uses the machine metaphor to make his point. For instance, he worries about how it is that an insect picks up pollen from a plant. Trying to simulate the activity himself, using a brush to poke around in the plant, he isn't very successful.

It then occurred to me that an insect in backing out of the flower would naturally push with some part of its body against the blunt and projecting upper end of the anther which overhangs the stigmatic surface. Accordingly I so held the brush that, whilst brushing upwards against the rostellum, I pushed against the blunt solid end of the anther . . . ; this at once eased the pollinia, and they were withdrawn in an entire state. At last I understood the *mechanism* of the flower. (Darwin 1862, 100)

What is fascinating is that Darwin uses the metaphor to explain aspects of nature that are machinelike in distinctive or peculiar ways.

In particular, he makes the point that nature has to make do with what it has rather than with what it would like to have. Hence, often, contrivances come out as though they were designed by (what the English would recognize as coming from) Heath Robinson or (what the Americans would recognize as coming from) Rube Goldberg.

> Although an organ may not have been originally formed for some special purpose, if it now serves for this end we are justified in saying that it is specially contrived for it. On the same principle, if a man were to make a *machine* for some special purpose, but were to use old wheels, springs, and pulleys, only slightly altered, the whole *machine*, with all its parts, might be said to be specially contrived for that purpose. Thus throughout nature almost every part of each living being has probably served, in a slightly modified condition, for diverse purposes, and has acted in the living *machinery* of many ancient and distinct specific forms. (Darwin 1862, 348)

KINDS OF MECHANISM

Darwin was deeply committed to the machine metaphor in the sense of artifact (mechanism 1) – the root metaphor of design. And this kind of thinking has persisted to the present. Indeed, it is the mark of the Darwinian. One of today's most influential thinkers, George Williams, has been explicit on this point.

> Whenever I believe that an effect is produced as the function of an adaptation perfected by natural selection to serve that function, I will use terms appropriate to human artifice and conscious design. The designation of something as the means or *mechanism* for a certain goal or function or purpose will imply that the *machinery* involved was fashioned by selection for the goal attributed to it. When I do not believe that such a relationship exists I will avoid such terms and use words appropriate to fortuitous relationships such as cause and effect. (Williams 1966, 9)

The other half of Mayr's claim about the proximate cause–ultimate cause distinction was that, in the study of origins, ultimate-cause thinking persists and predominates. We do still think in terms of final cause. And it would seem that he was right. Evolutionary biologists working in the Darwinian mode – and this is the mode in which such biologists almost all do work – think of organisms as machines and ask end-directed questions, just as did Darwin. What was the purpose or function of the plates on the back of the dinosaur stegosaurus? (We shall learn the belief today is that they regulate the body temperature of the brute.) What is the purpose or function of the big beaks possessed by certain species of Darwin's finch? (To crack hard nuts and seeds.) What is the purpose or function of the bright colors of the flowers of so many plants? (To attract pollinators, mainly birds and insects.)

It would be a mistake to think that the division between proximate and ultimate is sharp and decisive. All biologists think at times in terms of final cause; they use the artifact metaphor. Consider someone like Thomas Henry Huxley – who, as the archetypical agnostic, wanted nothing to do with God or with ideas of the world as His creation, and who was reducing everything to the blind workings of law, leading in a deterministic fashion to their inevitable ends: "the whole world, living and not living, is the result of the mutual interaction, according to definite laws, of the forces possessed by the molecules of which the primitive nebulosity of the universe was composed." What we cannot now say is that the end to which all of this leads has any purpose. The universe works like a clock, but we have no right to say that the purpose is telling time, any more than we can say that the purpose is senseless ticking. In this world, the forces that control all of this – including all of this in the world of organisms – are mechanisms. "And there seems to be no reply to this inquiry, any more than to the further, not irrational, question, why trouble oneself about matters which are out of reach, when the working of the *mechanism* itself, which is of infinite practical importance, affords scope for all our energies?" (Huxley 1873, 272–4)

Huxley (like Darwin) would seem to be completely committed to the sense of mechanism as process (mechanism 2). Yet (again like Darwin) he was quite committed to the notion that organisms in some sense are more than mere processes, openly speaking of organisms as "living machines in action" (mechanism 1). There was no conflict here. We seek proximate causes but make sense of them in the context of ultimate causes. We look for the processes, entirely eschewing vital forces, but we understand these processes in terms of the functioning of artifact-like organisms. With the early twentieth-century British physiologist William Bayliss (1915), we can quote Claude Bernard: "Pour le physiologist et le medicin experimentateur, l'organisme vivant is n'est qu'une machine admirable, douee des proprietes les plus merveilleuses, mise en action a l'aide des mecanismes les plus complexes et les plus delicats." ("For the physiologist and the medical experimenter, the living organism is nothing but an admirable machine, endowed with the most marvelous properties, put in motion by the most complex and most delicate mechanisms.")

If vitalism is entirely conquered, then this reliance on both senses of mechanism, both proximate and ultimate causes, must work the other way also. The student of origins must show us that although organisms are artifact-like, they are brought into being through non-vitalistic forces. No more of Buffon's *moule intérieur*, or of Blumenbach's *Bildungtrieb*. And this raises the question of whether the force that makes for living artifacts, natural selection, is a process like any other? Is it a process that falls under mechanism 2? If it is, then the picture is complete.

Interestingly, but perhaps not that significantly, Darwin did not use the language of mechanism when talking of natural selection (Ruse 2005b). I am sure that he thought of it as a force like Newtonian gravity. He was very keen to show that it had *vera causa* status like Newton's force, and it played the same central role in his science as did Newton's force in his science (Ruse 1975). Contrary to Kant, Darwin had no doubt that there could be a Newton of the blade of grass, and he intended to take on that role for himself. But Darwin's philosophical mentors, like John F. W. Herschel (1830) and William Whewell (1837, 1840), did not use mechanical language outside physics – Whewell was not even prepared to use it for chemistry – and Darwin basically followed suit, just as he followed suit in speaking of final causes even though his understanding of them was very different from that of those before him. However, in the sense understood in our discussion, it does seem clear that he was thinking of selection in mechanistic (type 2) terms. At times, Darwin has to defend the use of the metaphor of natural selection, and he admits that his language is all rather anthropomorphic, inasmuch as it implies that the force of evolution is a power or an intelligent being. Here, he makes it very clear that he does think of the world, including the organic world, as running according to natural processes: "I mean by nature only the aggregate action and product of many natural laws, – and by laws only the ascertained sequence of events" (Darwin 1868, 6).

I do not know who first actually called natural selection a "mechanism," using this in the simple sense of a process working or functioning, that is, mechanism without restriction (mechanism 2). I would not be surprised to find that it was an American, although expectedly (given that he was an evangelical Presbyterian) Darwin's first great over-the-Atlantic supporter, the Harvard botanist Asa Gray, never uses this term of selection in his collected essays, *Darwiniana*. (Just as expectedly, Gray constantly likens contrivance to machinelike processes.) In 1897, in *The Survival of the Unlike*, Liberty Hyde Bailey refers to natural selection as a hypothesis about the "controlling process or factor in evolution" (57), although he does use the language of machinery to talk about heredity (64). However, in his survey of evolutionary theories of 1907, *Darwinism Today*, Vernon L. Kellogg tells us that he is after the "factors and mechanism of organic evolution" (iii), and at the end of the book – denying the worth of selection as he moves toward the language we are seeking – concludes that "Darwinism as the all-sufficient and even most important causo-mechanical factor in species-forming and hence as the sufficient explanation of descent, is discredited and cast down" (374).

Around 1900, the modern theory of heredity ("genetics") started to pick right up, and no one had any hesitation about using the full-blooded mechanism language for this. Then, in the 1930s, the fortunes of natural selection likewise started to improve, and with this the language of mechanism (mechanism 2) starts to flood in. Theodosius Dobzhansky, in his *Genetics and the Origin of Species*, tells his reader that "*mechanisms* that counter-act the mutation pressure are known to exist. Selection is one of them . . ." (Dobzhansky 1937, 37–8). And again: "In its essence, the theory of natural selection is primarily an attempt to give an account of the probable *mechanism* of the origin of the adaptations of the organisms to their environment, and only secondarily an attempt to explain evolution at large" (150). (I suspect that this reading of the *Origin* would be news to its author.) More negatively, we have Sewall Wright, who was such an influence on Dobzhansky: "That evolution involves nonadaptive differentiation to a large extent at the subspecies and even the species level is indicated by the kinds of differences by which such groups are actually distinguished by systematists." Apparently only when you start to get up to the subfamily or family level do you start to get adaptive difference. "The principal evolutionary *mechanism* in the origin of species must thus be an essentially nonadaptive one" (Wright 1932, 168–9). More positively, over in England, we find Julian Huxley defending selection against various kinds of saltationism (evolution by macromutations) and pointing out that the "difference lies in the intermediary steps: in the one case the effect of use or function is supposed to be direct, in the other indirect, *via* the sifting *mechanism* of selection" (Huxley 1942, 39).

Finally, jumping to the present, let us turn to the first page of today's standard text on evolution, *Evolutionary Analysis*, by Scott Freeman and Jon C. Herron.

> Where did Earth's organisms come from? Why are there so many different kinds? How did they come to be so apparently well-designed to live where they live and do what they do? These are the fundamental questions of evolutionary biology. The answers are found in both the pattern and the *mechanism* of evolution. The pattern of evolution is descent with modification from common ancestors. The principle *mechanism* that drives this change is natural selection. (Freeman and Herron 2004, 1)

So what do we conclude? Obviously, that in evolutionary biology as in the rest of biology we have both proximate- and ultimate-cause type thinking. Mayr was right in making a distinction between kinds of explanation, wrong inasmuch as he suggested that physiologists use one kind of explanation exclusively and evolutionists another exclusively. Organisms are different, they are like artifacts. Rocks and rivers are not. Hence, in evolutionary biology, but not in physics or chemistry, it is appropriate to think in terms of the artifact-machine metaphor (mechanism 1). But the processes are processes of a kind that we find in the rest of science. They are mechanistic processes in the sense that we find in the physical sciences (mechanism 2). And what we find is that evolutionary biologists, just like other biologists, generally think in both senses at the same time when they are doing their work. For instance, the American evolutionist David Reznick argued that fish that are heavily predated as adults will mature quickly, so that their reproduction can occur before they are killed and eaten. This is an adaptation to a specific situation. Fish that are eaten with less discrimination, and especially fish that are eaten young, will have no such selective pressure driving them this way. With great skill and subtlety, Reznick was able to show that precisely these results hold for various little fish to be found in rivers on the island of Trinidad. Comparing two specific locations (high-predation patches were called generically "Chrenicichla" localities, and low-predation patches were called "Rivulus" localities), Reznick employed what is called the "mark-recapture" technique, whereby one simply catches all of the fish (in his case, guppies) in a particular location, marks them in some definite way so that they will not be confused with others, and then releases them back. Later one catches them all – at least one catches all of those remaining – and thus one can do comparative checks on the rates of predation. One can actually see whether the prey is indeed being eaten by the predators (Reznick et al. 1996). "If the differences in predation caused differences in guppy mortality rates, then the recapture probabilities of guppies from Crenicichla localities should be lower. If the Crenicichla prey selectively on adults, then the difference in mortality should be more dramatic for the adult age classes. We did, indeed, find that the overall probability of recapture was substantially and significantly lower in Crenicichla localities, implying higher mortality rates" (Reznick and Travis 1996, 269). As Reznick somewhat triumphantly concludes: "Such a result reveals a potential *mechanism* of life history evolution and thus goes a step farther in arguing that the differences in guppy life histories among Crenicichla and Rivulus localities represent an *adaptation* to predator induced mortality." A mechanism (sense 2) makes for an adaptation (mechanism sense 1).

DAWKINS REDUX

We can bring this chapter to a close. The machine metaphor functions in a different way in the life sciences than it does in the physical sciences. But the metaphor itself has conquered in the life sciences, just as it did in the physical sciences in earlier years. We can end, as we began, with the words of Richard Dawkins:

> Survival machines began as passive receptacles for the genes, providing little more than walls to protect them from the chemical warfare of their rivals and the ravages of accidental molecular bombardment. In the early days they 'fed' on organic molecules freely available in the soup. This easy life came to an end when the organic food in the soup, which had been slowly built up under the energetic influence of centuries of sunlight, was all used up. A major branch of survival machines, now called plants, started to use sunlight directly themselves to build up complex molecules from simple ones, re-enacting at much higher speed the synthetic processes of the original soup. Another branch, now known as animals, 'discovered' how to exploit the chemical labours of the

plants, either by eating them or by eating other animals. Both main branches of survival machines evolved more and more ingenious tricks to increase their efficiency in their various ways of life and new ways of life were continually being opened up. Sub-branches and sub-sub-branches evolved, each one excelling in a particular specialized way of making a living: in the sea, on the ground, in the air, underground, up trees, inside other living bodies. This sub-branching has given rise to the immense diversity of animals and plants which so impresses us today. (Dawkins 1976, 49)

FOUR

THINKING MACHINES

Scientists have been trying to determine why people need sleep for more than 100 years. They have not learned much more than what every new parent quickly finds out: sleep loss makes you more reckless, more emotionally fragile, less able to concentrate and almost certainly more vulnerable to infection. They know, too, that some people get by on as few as three hours a night, even less, and that there are hearty souls who have stayed up for more than week without significant health problems.

Now, a small group of neuroscientists is arguing that at least one vital function of sleep is bound up with learning and memory. A cascade of new findings, in animals and humans, suggest that sleep plays a critical role in flagging and storing important memories, both intellectual and physical, and perhaps in seeing subtle connections that were invisible during waking – a new way to solve a math or Easter egg problem, even an unseen pattern causing stress in a marriage.

The theory is controversial, and some scientists insist that it's still far from clear whether the sleeping brain can do anything with memories that the waking brain doesn't also do, in moments of quiet contemplation.

Yet the new research underscores a vast transformation in the way scientists have come to understand the sleeping brain. Once seen as a blank screen, a metaphor for death, it has emerged as an active, purposeful machine, a secretive intelligence that comes out at night to play – and to work – during periods of dreaming and during the netherworld chasms known as deep sleep.

Carey 2007

The title of this *New York Times* science section article was, not surprisingly, "An Active, Purposeful Machine That Comes Out at Night to Play."

THE SEVENTEENTH CENTURY

Let us return to Descartes. For all he wanted to make organisms into machines, and this would certainly have extended to human brains, when it came to thinking he wanted no part of it. Thought is a different substance. The clock metaphor applies to inanimate objects. It applies to plants and animals, even the higher ones. When it comes to humans it does not apply. Why not? Well, we know that we are thinking beings. *Cogito ergo sum.* Indeed, we know more than this. We know that our thinking is in some way a substance, something "whose whole essence or nature is only to think." We can conceive of a physical body without thought. We can also conceive of thinking without a physical body. Hence: "it is certain that this 'I' – that is to say, my soul, by virtue of which I am what I am – is entirely and truly distinct from my body and that it can be or exist without it" (Descartes 1964, 132).

This is all a bit fast, even for Descartes, and so a number of arguments are offered to show why mind and body are different substances. To give an example, in the Sixth Meditation, from which this quotation is taken, the argument depends on the indivisibility of the mind as opposed to the body. You can cut the body up into pieces. You cannot do the same with the mind. "For in truth, when I consider the mind, that is, when I consider myself in so far only as I am a thinking thing, I can distinguish in myself no parts, but I very clearly discern that I am somewhat absolutely one and entire. . . . But quite the opposite holds in corporeal or extended things; for I cannot imagine any one of them (how small soever it may be), which I cannot easily sunder in thought, and which, therefore, I do not know to be divisible" (p. 139). From this it follows that mind or soul is something quite different from body.

Cartesian "dualism" has had a magnetic hold on many people and indeed often is taken to be the commonsensical position. It was hardly new to Descartes. It is the philosophy of Plato, especially in works like the *Phaedo*, where Socrates talks of immortality. The survival after death is of the mind, not of the body, just as in the *Meno* the mind is remembering things that it had before the body was created. But, attractive though it may be – and even in recent times there have been serious thinkers who accept it (we will meet some later) – there are obviously big problems with mind-body dualism. Above all, there is the difficulty of seeing how mind and body could interact in any way. Unless you have God doing everything by way of linking up events in one world with events in the other – and this does seem to be the position of Descartes – it is not obvious why (say) a red hot poker pressed against the arm would cause a searing pain that affects the mind. Dogs and cats, which apparently have no minds, are likewise adverse to red hot pokers on their limbs and go through the motions to avoid them, exhibiting distress when they are burned. Why shouldn't not that be enough for us? Why should the mind get involved?

Or conversely (Descartes was not having the final or even the only word). The English political philosopher Thomas Hobbes was a materialist, and he thought that mind itself can be reduced to the material or physical. To most, this was simply implausible to the point of silliness, apart from having very dangerous theological implications. Given the times, Hobbes had to tread carefully, but he was not one for immortal souls. More balanced was the philosopher John Locke, who (in a very English fashion) was having nothing of the continental reduction of animals to machines. They may be that, but they are certainly more. "Brutes abstract not, yet are nor bare machines." Animals are not into abstract ideas or universals, but they can certainly reason in a primitive fashion, and this is a step beyond machines. "For if they have any ideas at all, and are not bare machines, (as some would have them,) we cannot deny them to have some reason. It seems as evident to me, that they do reason, as that they have sense; but it is only in particular ideas, just as they received them from their senses" (Locke 1689, Book 2, Chapter XI, section 11).

One could try to combine body and mind, the aim of the Dutch philosopher Benedict de Spinoza. He was thrown out of his synagogue at an early age, and reading his works one has a certain understanding of, if not sympathy for, the elders. Central to his philosophy (the *Ethics* especially) is a thoroughgoing naturalism, and this includes humans – every aspect of us. "The laws and rules of nature, according to which all things happen, and change from one form to another, are always and everywhere the same. So the way of understanding the nature of anything, of whatever kind, must also be the same, viz. through the universal laws of nature" (Spinoza 1985, Preface to Part III). Our thought is natural, and so are our actions and their motivations. As is well known, this does not mean that Spinoza was therefore a crude materialist. Anything but! He identified reality with God, a necessary being, Who has all of the attributes infinitely. Since thought and extension are two attributes, God has both – famously, Spinoza spoke of "God or nature" – which in some sense are different sides to the same thing. It is not so much a question of two substances, but of one substance seen in different ways. "A circle existing in nature and the idea of the existing circle, which is also in God, are one and the same thing, which is explained through different attributes. Therefore, whether we conceive nature under the attribute of Extension, or under the attribute of Thought, or under any other attribute, we shall find one and the same order, or one and the same connection of causes, i.e., that the same things follow one another" (Spinoza 1985, Proposition VII).

This thinking applies also to humans, and hence body and mind are one – not two things interacting but one thing showing different aspects of itself. This is hardly a machine in any sense as we might understand it, but do note that, as part of his version of naturalism, Spinoza was adamant that humans do not stand outside the causal nexus. In his argument, a version of which we are going to see again in a moment in Hume, Spinoza therefore held that free will has nothing to do with breaking the causal chain. It is much more a question of motives coming from within rather than from outside the individual. Moreover, as part of his psychology, Spinoza wanted no part of Aristotelian final causes somehow affecting human actions. All such causes are in fact efficient causes. "What is called a final cause is nothing but a human appetite insofar as it is considered as a principle, or primary cause, of some thing. For example, when we say that habitation was the final cause of this or that house, surely we understand nothing but that a man, because he imagined the conveniences of domestic life, had an appetite to build a house. So habitation, insofar as it is considered as a final cause, is nothing more than this singular appetite. It is really an efficient cause, which is considered as a first cause, because men are commonly ignorant of the causes of their appetites" (Spinoza 1985, Preface to Part IV). Note that in his refusal to countenance final causes as things that bring on human actions, Spinoza was implicitly escaping from what today's philosophers call the "missing goal object problem." If perchance the house does not get built, then it is odd to appeal to its existence as a cause of action. However, if the motivation is the thought of the finished house, then its noncompletion is irrelevant to the logic of the causal situation (Ruse 2003).

DAVID HUME

From our perspective, the colossi were Hume and Kant, although there were others of significance. We have already encountered La Mettrie, who argued that the body is a machine fueled by food. The brain is part of that machine, and it gives rise to thought. There is therefore no reason to think that the thought, the mind or soul, is anything more than a manifestation of the brain machine at work. "To be a machine, to feel, to think, to know how to distinguish good from bad, as well as blue from yellow, in a word, to be born with an intelligence and a sure moral instinct, and to be but an animal, are therefore characters which are no more contradictory, than to be an ape or a parrot and to be able to give oneself pleasure." From this it follows: "I believe that thought is so little incompatible with organized matter, that it seems to be one of its properties on a par with electricity, the faculty of motion, impenetrability, extension, etc" (La Mettrie 1912, 143–4). The target is Descartes and his identification of material substance with extension and of thinking substance with thought.

Bypassing others, people like David Hartley (1705–1757), who tried to reduce sensation and consequent thinking (via association of ideas) to physical vibrations starting in the nerves and then moving to the brain, let us jump straight to the great Scottish philosopher David Hume. Although there is much discussion about how strong are the connections and intentions, it is clear that in some sense Hume set out to do in philosophy what Newton did in science. He thought that his identification of the significance of the association of ideas – how we go from one idea to another through resemblance, contiguity or cause and effect – to be on a par with Newton's discovery of the law of gravitational attraction. We are therefore primed for an approach to the human mind that is going to be mechanistic, in some sense thinking of something – the brain or the mind or both – as a machine. And we are not disappointed.

The body is a machine:

> Thus, for instance, in the human body, when the usual symptoms of health or sickness disappoint our expectation; when medicines operate not with their wonted powers; when irregular events follow from any particular cause; the philosopher and physician are not surprised at the matter, nor are ever tempted to deny, in general, the necessity and uniformity of those principles by which the animal economy is conducted. They know that a human body is a mighty complicated machine: That many secret powers lurk in it, which are altogether beyond our comprehension: That to us it must often appear very uncertain in its operations: And that therefore the irregular events, which outwardly discover themselves, can be no proof that the laws of nature are not observed with the greatest regularity in its internal operations and government. (Hume 1777, section VIII, part 1, para. 67)

Reason is to a great degree a matter of instinct, certainly our reasoning about cause and effect is such, and this is all mechanical:

> I shall add, for a further confirmation of the foregoing theory, that, as this operation of the mind, by which we infer like effects from like causes, and *vice versa*, is so essential to the subsistence of all human creatures, it is not probable, that it could be trusted to the fallacious deductions of our reason, which is slow in its operations; appears not, in any degree, during the first years of infancy; and at best is, in every age and period of human life, extremely liable to error and mistake. It is more conformable to the ordinary wisdom of nature to secure so necessary an act of the mind, by some instinct or mechanical tendency, which may be infallible in its operations, may discover itself at the first appearance of life and thought, and may be independent of all the laboured deductions of the understanding. (Hume 1777, section V, part 2, para. 45)

Hume drew an analogy between the use of the physical body and the use of the mental body. "As nature has taught us the use of our limbs, without giving us the knowledge of the muscles and nerves, by which they are actuated; so has she implanted in us an instinct, which carries forward the thought in a correspondent course to that which she has established among external objects; though we are ignorant of those powers and forces, on which this regular course and succession of objects totally depends" (ibid.). Expectedly, this all put Hume with Locke and against Descartes when it came to the brutes. Instinct is the key to their thinking, and much the same holds for us. We are basically at one with the brutes: "Though the instinct be different, yet still it is an instinct, which teaches a man to avoid the fire; as much as that, which teaches a bird, with such exactness, the art of incubation, and the whole economy and order of its nursery" (section IX, para. 85).

It is thought that Hume started with his ideas about moral behavior and then worked back to knowledge and its psychological backing to explain such behavior. There is no surprise, therefore, that Hume's naturalistic way of thinking was tied in to his moral philosophy. For Hume, there is no will of God or any such thing standing behind moral beliefs. Rather, they are rooted in sentiments or emotions. This is not just a question of "if it feels ok, then it is ok," but rather of the feelings that stem from our human nature, which is designed to make us functioning members of family and society. (I use the language of design. Hume was very critical of the traditional design argument for God's existence, although he allowed that there might be something to it. He was not an evolutionist but would surely have welcomed an evolutionary explanation of human nature.) Most importantly, for Hume, it is the moral sentiment that influences reason and not conversely. Famously, he says that "reason is the slave of the passions," that is, "reason alone can never be a motive to any action of the will" (Hume 1739, Book 2, part 3, section 3, p. 266).

Finally, as with Spinoza, there is the denial of any freedom outside the normal causal chains. Hume is the spokesman for the standard "compatibilist" theory of free will, namely, that not only are freedom and causality possible together, but that without their being together freedom degenerates into incoherence. It is all a matter of what we mean by freedom. For Hume, as with Spinoza, it means freedom from external constraint, or "liberty." "By liberty, then we can only mean a power of acting or not acting, according to the determinations of the will; that is, if we choose to remain at rest, we may; if we choose to move, we also may. Now this hypothetical liberty is universally allowed to belong to every one who is not a prisoner and in chains" (Hume 1777, section VIII, part 1, para. 73). Therefore, a person who behaves according to Humean psychology – driven by the passions, guided by the reason, much of which is instinct – can be perfectly free. It is all a matter of constraint or not. This is certainly not to say that human behavior is outside causal necessity (in Hume's sense of obeying regularities). "Thus it appears, not only that the conjunction between motives and voluntary actions is as regular and uniform as that between the cause and effect in any part of nature; but also that this regular conjunction has been universally acknowledged among mankind, and has never been the subject of dispute, either in philosophy or common life" (Hume 1777, section VIII, part 1, para. 69). That moral actions have causes is not a weakness but a necessity. "Moral evidence is nothing but a conclusion concerning the actions of men, deriv'd from the consideration of their motives, temper and situation" (Hume 1739, Book 2, part 3, section 1, p. 260). Without our being able to infer people's thinking and actions in this way, society would simply break right down.

IMMANUEL KANT

Expectedly, the great German philosopher Immanuel Kant wanted little to do with this kind of philosophy, although much that he wrote was stimulated by the Scotsman's thinking. Kant's major problem is that of escaping the skepticism to which he saw Hume leading. If, as Hume argued, we can see no logical or other basis for our beliefs in the regularity of cause and effect and the persistence of objects and so forth, then the only escape – as Hume noted – lies in our psychology. There is no proof, but human nature lets us overcome this. "Most fortunately it happens, that since reason is incapable of dispelling these clouds, nature herself suffices to that purpose, and cures me of this philosophical melancholy and delirium, either by relaxing this bent of mind, or by some avocation, and lively impression of my senses, which obliterate all these chimeras. I dine, I play a game of back-gammon, I converse, and am merry with my friends; and when after three or four hours' amusement, I would return to these speculations, they appear so cold, and strain'd, and ridiculous, that I cannot find in my heart to enter into them any farther" (Hume 1739, Book 1, part 4, section 7, p. 175).

This was unacceptable to Kant, and his "Copernican revolution" was that of seeing that the necessity is put into nature by us rather than read straight from nature. He supposed that there are what he called "synthetic a priori" truths – truths that apply to the world (hence they are not analytic) but that are not derived from the world (hence they are a priori). Mathematics was his paradigm example of the synthetic a priori, but so also are the kinds of beliefs necessary to generate something like Newtonian mechanics. Hence, whereas for Hume cause and effect is simply a matter of our observing constant conjunction and then of our psychology making us think in terms of necessity, for Kant cause and effect are necessary conditions for thinking at all. In an important way, Kant is putting the mind outside the mechanical-machine picture. The empirical world of experience is machinelike, as we see and interpret it. But, although obviously humans are part of the empirical world, the mind is more a condition of seeing the world in this way than simply part of the machine.

We see this philosophy, which is notoriously difficult to understand, in Kant's discussion of the mind itself, and in his belief that it is a genuine unity rather than, as Hume would have it, simply a succession of impressions and ideas. Simple materialism or naturalism will not do, because there has to be "a transcendental ground of the unity of consciousness in the synthesis of the manifold of all our intuitions," within which all thinking takes place.

> This original and transcendental condition is no other than *transcendental apperception*. Consciousness of self according to the determinations of our state in inner perception is merely empirical, and always changing. No fixed and abiding self can present itself in this flux of inner appearances. Such consciousness is usually named *inner sense*, or *empirical apperception*. What has *necessarily* to be represented as numerically identical cannot be thought as such through empirical data. To render such a transcendental presupposition valid, there must be a condition which precedes all experience, and which makes experience itself possible. (Kant 1781, 136)

Reason, therefore, in Kant's philosophy has a very different role and status than it has in Hume's. Far from being instinct or something like that, it is the very condition of thought at all. This view carries over to Kant's moral thinking. Of course, as animals we are part of the physical world, but as humans we are more. Morality for Kant is a form of reason, namely, the conditions for living together as rational beings. This leads Kant to formulate his ultimate moral principle, the so-called Categorical Imperative – something like the epistemological necessities, a synthetic a priori truth. The Imperative is something with various forms, for instance, "Act always so that your actions could be universalized," something that makes lying immoral because if we all lied all of the time, then society would simply break down. Another popular formulation is "Treat humans as ends and not as means" – executing a couple of parking offenders would undoubtedly serve as warnings to the rest, but would be unfair to the victims, who would be punished excessively for the sake of others.

Predictably, Kant had little sympathy for the compatibilist position: "When I say of a man who commits a theft that, by the law of causality, this deed is a necessary result of the determining causes in preceding time, then it was impossible that it could not have happened; how then can the judgement, according to the moral law, make any change, and suppose that it could have been omitted, because the law says that it ought to have been omitted; that is, how can a man be called quite free at the same moment, and with respect to the same action in which he is subject to an inevitable physical necessity?" Fancy definitions of freedom that go along with causal necessity are no more than fancy definitions. "This is a wretched subterfuge with which some persons still let themselves be put off, and so think they have solved, with a petty word-jugglery, that difficult problem, at the solution of which centuries have laboured in vain, and which can therefore scarcely be found so completely on the surface." (Kant 1788, Book I, Chapter 3, para. 46)

Kant does not want to argue that, as free beings, we operate outside law, but rather – again a major break with Hume – that we operate according to self-imposed laws (hence we are "autonomous") that we recognize through reason. So, for instance, if I refuse to tell a lie even though it will lead to my own death, I am acting freely because I am acting according to the Categorical Imperative, something that I – not some external authority – impose upon myself, and the imposition comes not through instinct or anything psychological but through my reason recognizing the validity – the synthetic a priori necessity – of the Imperative. Invoking the already-encountered distinction between what Kant called the "phenomenal world," the world of physical experience, and the "noumenal world," the world of being in itself (the *Ding an sich*), freedom is a tricky concept that straddles the two worlds. We are balancing "the paradoxical demand to regard oneself qua subject of freedom as a noumenon, and at the same time from the point of view of physical nature as a phenomenon in one's own empirical consciousness" (Kant 1788, Preface, para. 11). In a sense, we are having our cake and eating it too.

Although we are now dealing with humans rather than organisms in general, in a way we are repeating some of the themes of the last chapter. Notwithstanding his great respect for the mechanical model in the physical world, Kant is not prepared to give the model free reign in the organic world (the last chapter) or in the human world (this chapter). Putting matters in the language of the last chapter, Kant insists on mechanism as process (mechanism 2) for the inorganic world. He insists on mechanism as artifact (mechanism 1) for the living world, and one can argue about whether he thought (as a Darwinian would think) that the processes of biology that bring about artifact-like entities (organisms) fall entirely under mechanism 1, or if you really need vital forces as well. When it comes to the human world, mechanism of any kind fails.

CHARLES DARWIN

Once again, Kant has drawn a line in the sand. Before, it was the claim that there can be no Newton of biology. Now, it is the claim that there can be no mechanical model of the mind. Once again, let us see how the world reacted to this challenge, beginning as before with the thinking of the great evolutionist, Charles Darwin. The coming of evolution led to major (if qualified) triumphs by the mechanical model in the organic world. What about the world of minds? What about the world of humans (and those animals that may share consciousness with us)? Charles Darwin certainly wanted a purely naturalistic account of humankind. From his days on HMS *Beagle*, when he had encountered the natives of Tierra del Fuego at the bottom of South America, he was absolutely convinced that we humans are animals like the rest of them and hence that we function like other organisms: we too are machines. Semihumorously, in his *Autobiography*, regretting his inability to enjoy music and poetry as he once did, he wrote: "My mind seems to have become a kind of machine for grinding general laws out of large collections of facts, but why this should have caused the atrophy of that part of the brain alone, on which the higher tastes depend, I cannot conceive" (Darwin 1958, 139).

It was in his *Descent of Man*, published in 1871, twelve years after the *Origin*, that Darwin gave his full account of human evolution, and there he makes it very clear that he thinks of humans as part of the animal world and that, however wonderful our powers of thought may be, they do not call for non-natural or nonmechanical causes. But digging back, the best place to discover Darwin's views on mind, brain, and thinking is the collection of private notebooks that he kept in the two or three years after he returned from the *Beagle* voyage in 1836. Here there is no need for caution or carefully phrased paragraphs, as he makes no bones about his materialism: "Thought (or desires more properly) being heredetary – it is difficult to imagine it anything but structure of brain heredetary, analogy points out to this. – love of the deity effect of organization. oh you Materialist!" (Barrett et al. 1987, C 166; the creative spelling is Darwin's). Nor is there any question about the relationship between brain and mind. The latter is a function of the former. "Why is thought, being a secretion of brain, more wonderful than gravity a property of matter? It is our arrogance, it our admiration of ourselves. –" (C 166) And in another notebook: "To study Metaphysics, as they have always been studied appears to me to be like puzzling at astronomy without mechanics. – Experience shows the problem of the mind cannot be solved by attacking the citadel itself. – the mind is function of body. – we must bring some *stable* foundation to argue from. –" (N 5)

As a corollary to this philosophy, Darwin is stone-cold certain that thinking is an adaptation and, as such, a candidate for natural selection. Indeed, the very first explicit intimation of natural selection as the cause of evolution – coming a month or two after he discovered it – focuses explicitly on the mind as adaptive. "An habitual action must some way affect the brain in a manner which can be transmitted. – this is analogous to a blacksmith having children with strong arms. – The other principle of those children which *chance* produced with strong arms, outliving the weaker ones, may be applicable to the formation of instincts, independently of habits." (N 42)

Expectedly, free will gets short shrift.

> With respect to free will, seeing a puppy playing cannot doubt that they have free will, if so all animals, then an oyster has & a polype (& a plant in some senses, perhaps, though from not having pain or pleasure, actions unavoidable & only to be changed by habits). Now free will of oyster, one can fancy to be direct effect of organization, by the capacities its senses give it of pain or pleasure. If so free will is to mind, what chance is to matter /M. Le Compte/ – the free will (if so called) makes change in bodily organization of oyster, so may free will make change in man. – the real argument fixes on heredetary disposition & instincts. – Put

it so. – Probably some error in argument, should be grateful if it were pointed out. My wish to improve my temper, what does it arise from, but organization, that organization may have been affected by circumstances & education & by the choice which at that time organization gave me to will – Verily the faults of the fathers, corporeal & bodily, are visited upon the children. – (M 72–3)

Note that, when Darwin talks of "chance," he always means "according to law, of which we are ignorant."

One is tempted to believe phrenologists are right about habitual exercise of the mind, altering form of the head, & thus these qualities become heredetary. – When a man says I will improve my powers of imagination, & does so. – is not this free will,– he improves the faculty according to the usual method, but what urges him, – absolute free will, motive may be anything ambition, avarice, &c&c An animal improves because its appetites urges it to certain actions, which are modified by circumstances, & thus the appetites themselves become changed. – appetites urge the man, but indefinitely, he chooses (but what makes him fix!?) – frame of mind, though perhaps he chooses wrongly, – & what is frame of mind owing to. –) – I verily believe free-will and chance are synonymous. – Shake ten thousand grains of sand together & one will be uppermost:– so in thoughts, one will rise according to law. (M 30)

In other words, for Darwin, free will is a matter of mental laws of which we are (or may be) ignorant. It is not something that takes us outside the machine.

Finally, what about reason and morality and so forth, the sorts of things of concern to Hume and Kant? As far as they are concerned, Darwin is very much in the tradition of Hume, save that he gives an evolutionary backing to everything. "Origin of man now proved. – Metaphysic must flourish. – he who understands baboon would do more towards metaphysics than Locke" (M 84e, August 16, 1838 – that is, before the discovery of selection). Then a few days later, he lets us see explicitly the line that he is following. Instinct formed by evolution gives us innate knowledge. "Plato says in Phaedo that our *"necessary ideas"* arise from the preexistence of the soul, are not derivable from experience. – read monkeys for preexistence – " (M 128, September 4, 1838 – Darwin makes it clear in the notebook that he is relying here on his older brother, another Erasmus Darwin, for the information about Plato).

Darwin was a scientist, not a philosopher, and he does not follow through on these speculations. But he does return one more time to the topic, a couple of years later in 1840. He is reading an article by John Stuart Mill in the *Westminster Review*. Mill is explaining the difference between the empiricists like Locke, who think that all knowledge is based on experience, and those like Kant, who think that the mind structures experience. "Every consistent scheme of philosophy requires, as its starting point, a theory respecting the sources of human knowledge.... The prevailing theory in the eighteenth century was that proclaimed by Locke, and attributed to Aristotle – that all our knowledge consists of generalizations from experience.... From this doctrine Coleridge with... Kant... strongly dissents.... He distinguishes in the human intellect two faculties... Understanding and Reason. The former faculty judges of phenomena, or the appearance of things, and forms generalizations from these: to the latter it belongs, by direct intuition, to perceive things, and recognize truths, not cognizable by our senses." Mill makes it very clear that he himself is in the empiricist camp. "We see no ground for believing that anything can be the object of our knowledge except our experience, and what can be inferred from our experience by the analogies of experience itself."

Darwin reasserts his evolutionary position:

> Westminster Review, March 1840 p. 267 – says the great division amongst metaphysicians – the school of Locke, Bentham, & Hartley, &. and the school of Kant. to Coleridge, is regarding the sources of knowledge. – whether "anything can be the object of our knowledge except our experience". – is this not almost a question whether we have any instincts, or rather the amount of our instincts – surely in animals according to the usual definition, there is much knowledge without experience. so there *may* be in men – which the reviewer seems to doubt. (OUN, 33)

Morality gets a similarly naturalistic treatment – "I am tempted to say that those actions which have been found necessary for long generation (as friendship to fellow animals in social animals) are those which are good & consequently give pleasure" (M 132e). However, we can now turn profitably to the *Descent of Man* (1871), for it is there that he does give his most detailed analysis of morality.

What is morality? Darwin answered this question in much the same way it would have been answered by most of his countrymen of the day – he opted for some kind of utilitarianism: The Greatest Happiness for the Greatest Number. Admittedly, however, this was framed more in biological terms than in terms of purely human pleasures. "The term, general good, may be defined as the term by which the greatest possible number of individuals can be reared in full vigour and health, with all their faculties perfect, under the conditions to which they are exposed" (Darwin 1871, 1, 98). Darwin was fairly sophisticated in his analysis of morality. He recognized that we humans clearly have selfish or self-directed desires. But he saw that, without something more, such beings as humans cannot form societies. Although he was certainly never going to accept the metaphysics of Kant, perhaps his reading of the great philosopher (Darwin had read the *Metaphysics of Morals*) had helped him here. We have to have some sense or feeling of sociability – some felt need to be on good working terms with our fellow humans. Darwin recognized that there is a bit of a chicken-and-egg situation here. Are we social because we need to get together, or can we get together because we have the ability and need to be social? Either way, we can see the need for natural selection. Those who are social will do well, and those who are sourpusses will be singled out and isolated.

Darwin saw the need of some kind of moral sense to drive our relationships with others. Even though we have this in the most developed form, it is not unique to humans. Our moral or social sense is something possessed by the animals and so might have been expected to evolve. This is not to say that humans are just animals. Intelligence and the ability to be self-reflective are important here. Humans are fully moral – uniquely moral – because we have the ability to think about our actions, to judge them, to try to influence ourselves with respect to future behavior. In short, we have a con- science, capable of what we today might want to call second-order desires that sort through the first-order desires. At the first order, I want to look to myself, and as a social being I also want to look to others. I look exclusively to myself. Now I think about it, and am disgusted with myself and want not to have this feeling again. So the next time I try to do better.

If you suspect that this all continues to be rather Kantian- sounding, you are probably right. Darwin backed up his thinking with a purple-prose passage from the sage of Königsberg (Kant's home town). "Duty! Wondrous thought, that workest neither by fond insinuation, flattery, nor by any threat, but merely by holding up thy naked law in the soul, and so extorting for thyself always reverence, if not always obedience; before whom all appetites are dumb, however secretly they rebel; whence thy original?" (Darwin 1871, 1, 70). However, be careful not to see the influence as too great or pervasive. For Kant, morality is a necessary condition for rational beings living together. Darwin, in a very British empiricist sort of way, calmly suggested that, had evolution gone another way, we might be rational but with a very different kind of normative morality. We could have been like the worker bees, thinking that at times our greatest moral obligation is to kill our lazy, useless brothers! All in all, the discussion is very much in the style of Hume: "The aid which we feel impelled to give to the helpless is mainly an incidental result of the instinct of sympathy, which was originally acquired as part of the social instincts, but subsequently rendered...more tender and more widely diffused" (Darwin 1871, 1, 101).

What of the causes of morality? Why do we have a moral sense? At the very most, Darwin was on the cusp of biology and culture. He never took seriously the idea that we might have a basic selfish biological nature and that culture then laid morality on top of this. Biology is more important. Genuine moral urges are part of our heredity. However, here is a paradox. Darwin accepted fully that morality is a group characteristic. That is the whole point about morality. He accepted also that morality seems to go against individual interest. "It is extremely doubtful whether the offspring of the more sympathetic and benevolent parents, or of those who were the most faithful to their comrades, would be reared in greater numbers than the children of selfish and treacherous parents of the same tribe" (Darwin 1871, 1, 163). How is one to escape the dilemma? For a start, Darwin offered a version of a mechanism that has been called "reciprocal altruism." You scratch my back and I'll scratch yours (Trivers 1971). "In the first place, as the reasoning powers and foresight of the members became improved, each man would soon learn that if he aided his fellow-men, he would commonly receive aid in return. From this low motive he might acquire the habit of aiding his fellows; and the habit of performing benevolent actions certainly strengthens the feeling of sympathy which gives the first impulse to benevolent actions. Habits, moreover, followed during many generations probably tend to be inherited" (Darwin 1871, 1,

163–4). But Darwin also did allow that people are moved by the praise and condemnation of their fellows. If I do not have to worry about praise and blame, then I might act selfishly and have more offspring. It seems that it is only if the worry, which is something for the good of the group, can be promoted by selection for the benefit of the group that it gets preserved. Having said this, however, Darwin made it very clear that he is talking about the tribe and not the species, and he stressed that he saw the tribe as being interrelated. Ultimately, it is individual payoff that counts.

EVOLUTIONISTS AFTER DARWIN

Although many people became evolutionists after the *Origin*, natural selection was less favorably received. That had to wait until the second quarter of the twentieth century before it would come into its own. Darwin's self-appointed "bulldog," Thomas Henry Huxley, had little sympathy for selection, but he was committed to evolution as science and to naturalism as philosophy. For him, animals are automata, and the same holds true of humans. We are all machines, and basically consciousness is just irrelevant froth on the top. "But though we may see reason to disagree with Descartes' hypothesis that brutes are unconscious machines, it does not follow that he was wrong in regarding them as automata. They may be more or less conscious, sensitive, automata; and the view that they are such conscious machines is that which is implicitly, or explicitly, adopted by most persons" (Huxley 1874, 237–8). Basically, the same is true of humans. We are physical machines, and consciousness in some way sits on top of this.

> It may be assumed, then, that molecular changes in the brain are the causes of all the states of consciousness of brutes. Is there any evidence that these states of consciousness may, conversely, cause those molecular changes which give rise to muscular motion? I see no such evidence. The frog walks, hops, swims, and goes through his gymnastic performances quite as well without consciousness, and consequently without volition, as with it; and, if a frog, in his natural state, possesses anything corresponding with what we call volition, there is no reason to think that it is anything but a concomitant of the molecular changes in the brain which form part of the series involved in the production of motion. (239–40)

Brute consciousness, therefore, is a by-product with no powers of its own. It can no more bring about change or action than a steam whistle on a locomotive can affect the motions of the engine. The same is true of humans. "It is quite true that, to the best of my judgment, the argumentation which applies to brutes holds equally good of men; and, therefore, that all states of consciousness in us, as in them, are immediately caused by molecular changes of the brain-substance. It seems to me that in men, as in brutes, there is no proof that any state of consciousness is the cause of change in the motion of the matter of the organism" (243–4). Free will, then, is a kind of illusion. At most it is a matter of (in the Humean sense) not being constrained from outside. In no way is it a matter of escaping the web of causality.

Note that this consciousness-akin-to-the-whistle-on-a-locomotive position (it is known by professional philosophers as epiphenomenalism) is not at all Darwin's position. For him, it would be unthinkable to suppose that thought has no role in human life. Whatever the connection between mind and brain, Darwin clearly thinks of minds – thinking – as important in survival and reproduction. Brains exist to support minds in their activities. For Darwin, organic features exist because they are adaptations, because they help the organism in the struggle to survive and reproduce. One's first question, therefore, is: "What's it for?" Huxley never felt great enthusiasm for natural selection, and as an anatomist who was interested in tracing similarities ("homologies") between organisms, and whose specimens were usually found dead and preserved on a dissecting table (and hence not functioning), adaptation was not a major concern of his.

The difference between Darwin and Huxley is highlighted by a very pro-Darwinian American, the pragmatist philosopher and psychologist William James: "It is to my mind quite inconceivable that consciousness should have *nothing to do* with a business which it so faithfully attends. And the question, 'What has it to do?' is one which psychology has no right to 'surmount,' for it is her plain duty to consider it. The fact is that the whole question of interaction and influence between things is a metaphysical question, and cannot be discussed at all by those who are unwilling to go into matters thoroughly." Hence, "to urge the automaton-theory upon us, as it is now urged, on purely *a priori* and *quasi*-metaphysical grounds, is an *unwarrantable impertinence in the present state of psychology*"

(James 1880, 1, 138). For a Darwinian, this will never do: "It is very generally admitted, though the point would be hard to prove, that consciousness grows the more complex and intense the higher we rise in the animal kingdom. That of a man must exceed that of an oyster. From this point of view it seems an organ, superadded to the other organs which maintain the animal in the struggle for existence; and the presumption of course is that is helps him in some way in the struggle, just as they do. But it cannot help him without being in some way efficacious and influencing the course of his bodily history" (ibid.).

At the least, in James's thinking, consciousness helps us to choose between courses of action. The unguided brain can only offer the choices. Consciousness directs: "The brain is an instrument of possibilities, but of no certainties. But the consciousness, with its own ends present to it, and knowing also well which possibilities lead thereto and which away, will, if endowed with causal efficacy, reinforce the favorable possibilities and repress the unfavorable or indifferent ones. The nerve-currents, coursing through the cells and fibres, must in this case be supposed strengthened by the fact of their awaking one consciousness and dampening by awakening another" (1, 141–2). Another reason for consciousness is that pleasure and pain are good guides to action: "Starvation, suffocation, privation of food, drink and sleep, work when exhausted, burns, wounds, inflammation, the effects of poison, are as disagreeable as filling the hungry stomach, enjoying rest and sleep after fatigue, exercise after rest, and a sound skin and unbroken bones at all times, are pleasant." What is going on here? Obviously, we have urges to action: "if pleasures and pains have no efficacy, one does not see (without some such *à priori* rational harmony as would be scouted by the 'scientific' champions of the automaton-theory) why the most noxious acts, such as burning, might not give thrills of delight, and the most necessary ones, such as breathing, cause agony" (1, 144).

In short: "*A priori* analysis of both brain-action and conscious action shows us that if the latter were efficacious it would, by its selective emphasis, make amends for the indeterminateness of the former; whilst the study *a posteriori* of the *distribution* of consciousness shows it to be exactly such as we might expect in an organ added for the sake of steering a nervous system grown too complex to regulate itself. The conclusion that it is useful is, after all this, quite justifiable" (1, 144).

THE TWENTIETH CENTURY

If nothing else, for all their differences, Huxley and James both show how the coming of Darwinism confirmed beliefs that the mind is susceptible to naturalistic understanding and that a mechanical model is the right framework within which to work. There are others to whom we could turn for further illumination of how people wrestled with consciousness in the immediate post-Darwinian period. Included here would be people like Conwy Lloyd Morgan (1894), a student of Huxley, who studied behavior, eschewing final-cause explanations, and who tried to understand actions in terms of simple trial-and-error learning. However, the necessary point is made, and we can now move on to the twentieth century, when, as is well known, biology as applied to humans got eclipsed somewhat. The social sciences started to rise to prominence, and often there was little interest in thinking of our own species from an evolutionary perspective, let alone from a Darwinian perspective. If we were offering a standard history, one might feel it necessary to spend time exploring in some detail what this all meant from the viewpoint of the machine metaphor. However, we can legitimately move quickly on and down to the present. Although there were certainly those who did challenge the machine model as applied to the mind – for instance, existential psychoanalytic theory (as found in the thinking of such people as Viktor Frankl and Rollo May) – for the social sciences even more than for the biological sciences, the machine model generally was the starting point of inquiry. So basic was it that people did not even realize that they were committed to it.

Take the thinking of Freud, for instance. He is of course a highly controversial figure, and there are those who question whether he should be thought of as a scientist. People like the late Karl Popper were very critical of any pretentions on the part of Freud's kind of thinking to be considered genuine empirical study. Moreover, in many respects Freud was as much a biologist (albeit not much of a Darwinian) as a social scientist. But whatever the status of his ideas, that they have been incredibly influential, in culture if not in science and medicine, is beyond doubt. And that Freud was a hard-line proponent of a mechanical view of mind is a given. It is all a matter of nature's laws determining thought and action. Take his greatest work, *The Interpretation of Dreams*, where he writes of dreams as phenomena that give one kind of picture or story that must be interpreted and given another kind of picture or story, one that makes sense. "The dream-thoughts and the dream-content present themselves as two descriptions of the same content in two different languages; or, to put it more clearly, the dream-content appears to us as a translation of the dream-thoughts into another mode of expression, whose symbols and laws of composition we must learn by comparing the origin with the translation. The dream-thoughts we can understand without further trouble the moment we have ascertained them. The dream-content is, as it were, presented in hieroglyphics, whose symbols must be translated, one by one, into the language of the dream-thoughts" (Freud 1900, Chapter 6, "The Dream Work"). Obviously there is something mechanical going on here. We have concerns and desires and intentions, often without being fully aware of their nature, and these come bubbling up in the form of dreams. It is a process happening automatically, without our control, and Freud often uses the language of machines to describe the particulars of what is going on. As in: "The mechanism of dream-formation is favourable in the highest degree to only one of the logical relations. This relation is that of similarity, agreement, contiguity, just as; a relation which may be represented in our dreams, as no other can be, by the most varied expedients" (Freud 1900, "The Means of Representation in Dreams"). Elsewhere, Freud openly writes of this whole process as the "dream mechanism" (Chapter Two of *Dream Psychology: Psychoanalysis for Beginners*).

In that area of the social sciences proper that is devoted to the mind – psychology – research and speculation was dominated until well after the middle of the century by the approach toward which Lloyd Morgan was pointing, behaviorism. This comes in various stripes. Perhaps the weakest is some form of methodological behaviorism, favored by (among others) the American psychologist John Watson, arguing that research should confine itself to what can be observed and measured. There is no denial of mind itself, but it is argued that minds are beyond scientific inquiry and that study should be of, and only of, the behavior exhibited by animals and humans. Next there is psychological behaviorism, found in the work of people like Edward Thorndyke (1905) and B. F. Skinner (1938, 1953), something that tries to put methodological behaviorism to work, looking at the effects of stimuli and repeated patterns and so forth. Ivan Pavlov's (1927) work on his dogs, getting them to salivate given certain stimuli, was a classic case of psychological behaviorism.

Often this sort of approach was linked to a *tabula rasa* (blank slate) view of the mind, where all new information and abilities have to be built from the ground up, as it were. Finally and most strongly, there is analytic behaviorism, found in philosophers like Gilbert Ryle (1949) and the later Wittgenstein (and, a point to remember, possibly in Ryle's student, the philosopher Daniel Dennett), which simply identifies mental content with behaviors or dispositions to such behavior. This obviously is as much a metaphysical thesis as anything purely scientific.

Once again, one can hardly exaggerate the extent to which all three of these approaches exist within the machine metaphor. There are times when one almost feels one is back with Hobbes, or certainly with La Mettrie. Later, some of this kind of thinking will receive critical scrutiny. For now, let us move on quickly. The aim here is not really to get to the truth of such thinking, although it is worth noting that one of the reasons why behaviorism started to crumble was the work of the linguist Noam Chomsky, who, in the 1950s, presented solid evidence that children's language acquisition was far from a simple matter of responding to stimuli in the way that the behaviorists supposed. Chomsky's thesis, which suggests that our evolutionary past might be very important, is that children have capacities that far exceed the information that is given to them. The mind is not a blank slate waiting to be written upon by the adults that the child encounters.

Chomsky, notoriously, is not very keen on Darwinism, but those coming after him have been more in the traditional evolutionary paradigm. Therefore, without further ado let us now move straight to the work of Darwin's most direct heirs, those Darwinian evolutionists most interested in the mind. Since, as Darwin saw, this usually means being interested in social behavior, the people to whom we are referring are those known as "sociobiologists" – although today those specifically interested in human nature more commonly go under the title of "evolutionary psychologists."

EVOLUTIONARY PSYCHOLOGY

The starting point for the modern Darwinian attack on the mind and its thinking abilities is the work of the Harvard expert on the social insects Edward O. Wilson. In the 1970s, he extended his gaze to the whole of the animal world, and then refocused on humans, especially toward the end of his magisterial survey, *Sociobiology: The New Synthesis* (1975), and a spin-off book, *On Human Nature* (1978). The whole approach is thoroughly mechanistic, with the brain regarded as some kind of machine driving thought, something that in turn brings on the behavior of the individual human as he or she functions in groups. So decidedly mechanistic was the approach that critics accused Wilson of genetic determinism, seeing humans as mere marionettes, controlled by the unthinking DNA that codes for life. I am not sure that this criticism is entirely fair. Wilson's ideas certainly could accommodate a compatibilist view of free will that goes back to Hume and beyond. But it is true that the machine metaphor is dominant, perhaps sometimes to an uncomfortable degree. We see this right on the opening page of *On Human Nature*.

> How does the mind work, and beyond that why does it work in such a way and not another, and from these two considerations together, what is man's ultimate nature?
> We keep returning to the subject with a sense of hesitancy and even dread. For if the brain is a machine of ten billion nerve cells and the mind can somehow be explained as the summed activity of a finite number of chemical and electrical reactions, boundaries limit the human prospect – we are biological and our souls cannot fly free. If humankind evolved by Darwinian natural selection, genetic chance and environmental necessity, not God, made the species. (Wilson 1978, 1)

The answer to the question "How does the mind work?" seems to lie (in Darwinian fashion) in our past and in the inherited memories that it brings with it. All the evidence is "of a social world too complex to be constructed by random learning processes in a lifetime."

> So the human mind is not a tabula rasa, a clean slate on which experience draws intricate pictures with lines and dots. It is more accurately described as an autonomous decision-making instrument, an alert scanner of the environment that approaches certain kinds of choices and not others in the first place, then innately leans towards one option as opposed to others and urges the body into action according to a flexible schedule that shifts automatically and gradually from infancy into old age. The accumulation of old choices, the memory of them, the reflection on those to come, the re-experiencing of emotions by which they were engendered, all constitute the mind. (p. 67)

The only concession that Wilson allows is that things might just be too complex for us to follow at an individual level. "We note that even if the basis of mind is purely mechanistic, it is very unlikely that any intelligence could exist with the power to predict the precise actions of an individual human being, as we might to a limited degree chart the path of a coin or the flight of a honeybee. The mind is too complicated a structure, and human social relations affect its decisions in too intricate and variable a manner, for the detailed histories of individual human beings to be predicted in advance by the individuals affected or by other human beings" (p. 77). However, even if not at the individual level, Wilson is sure that humans are predictable en masse, so the mechanistic hypotheses certainly holds at this level. Materialism, mechanism, evolutionism – all is bound in one synthetic overall picture.

> The core of scientific materialism is the evolutionary epic. Let me repeat its minimum claims: that the laws of the physical sciences are consistent with those of the biological and social sciences and can be linked in chains of causal explanation; that life and mind have a physical basis; that the world as we know it has evolved from earlier worlds obedient to the same laws, and that the visible universe today is everywhere subject to these materialist explanations. (Wilson 1978, 201)

In the language of the previous chapter, Wilson is committed to seeing humans as artifacts (mechanism 1) run by natural processes (mechanism 2).

In the thirty years since Wilson penned these words, the science has galloped forward. But the philosophy remains the same. In the realm of knowledge and reason, the most important work has been that of John Tooby and Leda Cosmides (Barkow, Cosmides and Tooby 1991; Tooby, Cosmides and Barrett 2005). They have detailed how our thinking is governed by innate principles of reasoning – not necessarily those of strict formal logic, but the kinds of inference making that are important to us in everyday life. In discussing their work, the Harvard psychologist Steven Pinker (1997) makes this point clearly. He dismisses claims that (referring to a metaphor that will get more discussion in the next section) the mind is simply an all-purpose computer and endorses Cosmides's claim that reasoning is a social activity more often than not. Hence, an important aspect of the mind is as an instrument for detecting cheating. Consider the following puzzle (known as the "Wason selection task"). You are told that if a playing card has a D on one side, then it has a 3 on the other. Which of the following four cards must you turn over to check whether the rule is followed?

D F 3 7

Most people chose D and 3, whereas the correct answer is D and 7. But if presented with the same formal question in a familiar context, most people get the answer right. You are a bouncer for a bar, and the rule is that in order to be served a beer, you must be over eighteen. The choices now are: a beer drinker, a Coke drinker, a twenty-five-year-old, and a sixteen-year-old. It is "obvious" that you must check the IDs and/or drinks of the beer drinker and the sixteen-year-old. No one chooses to check the twenty-five-year-old.

Pinker concludes (in the language of machines): "The mind seems to have a cheater-detector with a logic of its own. When standard logic and cheater-detector logic coincide, people act like logicians; when they part company, people still look for cheaters" (Pinker 1997, 337). And natural selection lies behind this. "Any selfless behavior in the natural world needs a special explanation. One explanation is reciprocation: a creature can extend help in return for help expected in the future. But favor–trading is always vulnerable to cheating. For it to have evolved it must be accompanied by a cognitive apparatus that remembers who has taken and ensures that they give in return" (ibid.).

In the realm of moral behavior, evolutionary psychological thinking is just as embedded within the machine metaphor. Consider a recent discussion of morality – or, more specifically, of the "moral organ" used for making ethical decisions. Pinker's Harvard colleague Marc Hauser believes that morality functions in a way akin to language, and, using this analogy, he writes:

> We posit a theory of universal moral grammar which consists of the principles and parameters that are part and parcel of this biological endowment. Our universal moral grammar provides a toolkit for building possible moral systems. Which particular moral system emerges reflects details of the local environment or culture, and a process of environmental pruning whereby particular parameters are selected and set early in development. (Hauser 2006, 215)

I need hardly stress that tool kits are things owned and used by garage mechanics, as they plunge into the workings of the engine under the hood. And by computer repair people as they fix your hard drive. In other words, the things used by people who work on machines.

Similar kinds of mechanistic thinking dominate in work reported by the eminent ethicist Peter Singer. There is a famous paradox in moral philosophy known as the trolley problem. Suppose you see a runaway trolley that is on track to kill five unsuspecting people. Suppose also that you are beside a switch that would let you divert the trolley to a side line, where there is only one unsuspecting person at risk. Would you flip the switch? Most people say "Yes!" Now suppose that, rather than a switch, you are standing next to a fat man whom you could push in the trolley's way and save five people. Most people say "no!" (Nothing against fat people, but he has got to be big enough to stop the trolley, and you yourself are not, so the suicide option is ruled out.) Singer cites work showing that we use different parts of the brain to make these two judgments and that this is related to the fact that the pushing case involves a personal violation and raises the related emotions, whereas the switching case is more impersonal and more open to reflective reasoning.

> When people were asked to make judgments in the "personal" cases, parts of their brains associated with emotional activity were more active than when they were asked to make judgments in "impersonal" cases. More significantly, those who came to the conclusion that it would be right to act in ways that involve a personal violation, but minimize overall harm – for example, those who say that it would be right to push the stranger off the footbridge – took longer to form their judgment than those who said it would be wrong to do so.
>   When [experimenter Joshua C.] Greene looked more closely at the brain activity of these subjects who say "yes" to personal violations that minimize overall harm, he found that they show more activity in parts of the brain associated with cognitive activity than those that say "no" to such actions. (Singer 2005, 341–2)

Why is all of this? Singer relates it to our biology. We are prisoners of our past. Human evolution took place in small groups, where violence was an everyday phenomenon. If not within the group, then between groups of proto-humans. Those people who handled this kind of situation better than their fellows were more likely to survive and reproduce. They were more likely to develop and pass on adaptations for dealing with conflict situations, for dealing with interpersonal interactions, for dealing with points where split-second decisions are needed, and so forth. A lot of this will involve emotion rather than reason. Our would-be ancestors did not have the luxury of time to make calculations. And so when we are presented today with real-life situations, we will act one way, using emotion. When we are presented with hypothetical situations, we must turn to other capacities, often involving reason.

> The thought of pushing the stranger off the footbridge elicits... emotionally based responses. Throwing a switch that diverts a train that will hit someone bears no resemblance to anything likely to have happened in the circumstances in which we and our ancestors lived. Hence the thought of doing it does not elicit the same emotional response as pushing someone off a bridge. So the salient feature that explains our different intuitive judgments concerning the two cases is that the footbridge case is the kind of situation that was likely to arise during the eons of time over which we were evolving; whereas the standard trolley case describes a way of bringing about someone's death that has only been possible in the past century or two, a time far too short to have any impact on our inherited patterns of emotional response. (pp. 347–8)

I need hardly say that this whole discussion is taking place within a mechanistic framework – the brain acts in certain ways, and these influence our thinking and acting. Even the reasoning is something that is part of our inherited biology rather than a Kantian process that stands outside the causal framework. Human sociobiology or evolutionary psychology may not be genetically deterministic in the manner charged by the critics – the whole point of the trolley problem is that we have the ability to make choices – but machine thinking (in the pattern set by Wilson) predominates.

COGNITIVE SCIENCE

Evolutionary psychology sets the background. Our thought and behavior should be understood in Darwinian terms. But how is one to work out the specifics? How is one to peer into the actual ways in which decisions are made and actions taken? Here, famously, we encounter the area of inquiry committed to the biggest machine metaphor of them all. I refer of course to cognitive science and its base metaphor: the brain-mind is a computer. The physical side, the hardware, is the brain. Thinking is the mental side, the software. This is the theme stressed again and again, from introductory textbooks to sophisticated research papers. Consider the opening paragraphs of a popular and deservedly praised introduction, *Mindware*, by the philosopher Andy Clark.

> The computer scientist, Marvin Minsky once described the human brain as a meat machine – no more, no less. It is, to be sure, an ugly phrase. But it is also a striking image, a compact expression of both the genuine scientific excitement and the rather gung-ho materialism that has tended to characterize the early years of cognitive science research. Mindware – our thoughts, feelings, hopes, fears, beliefs, and intellect – is cast as nothing but the operation of the biological brain, the meat machine in our head. This notion of the brain as a meat *machine* is interesting, for it immediately invites us to focus not so much on the material (the meat) as on the machine: the way the material is organized and the kind of operation it supports. The same machine can, after all, often be made of iron, or steel, or tungsten, or whatever. What we confront is thus both a rejection of the idea of mind as immaterial spirit-stuff and an affirmation that mind is best studied from a kind of engineering perspective that reveals the nature of the machine that all that wet, white, gray, and sticky stuff happens to build.

Clark then goes straight on to say:

> What exactly is meant by casting the brain as a machine, albeit one made out of meat? There exists a historical trend, to be sure, of trying to understand the workings of the brain by analogy with various currently fashionable technologies: the telegraph, the steam engine, and the telephone switchboard are all said to have had their day in the sun. But the "meat machine" phrase is intended, it should now be clear, to do more than hint at some rough analogy. For with regard to the very special class of machines known as computers, the claim is that the brain (and, by not unproblematic extension, the mind) actually *is* some such device. It is not that the brain is somehow *like* a computer: everything is like something else in some respect or other. It is that neural tissues, synapses, cell assemblies, and all the rest are just nature's rather wet and sticky way of building a hunk of honest–to-God computing machinery. Mindware, it is then claimed, is found "in" the brain in just the way that software is found "in" the computing system that is running it. (Clark 2000, 7–8)

Well, yes – but. The brain is still literally a brain and still metaphorically a computer. Computers are the sorts of things that authors use to write books – things made out of plastic and copper and so forth. You can extend the term "computer" to include brains, but only if you decide that the similarities are so great as to warrant doing so. But you have to make the extension. In a way, though, this is not the truly important point. As we draw to the end of the historical part of this book, the truly important point worth stressing is that which was made as we launched into our history. Metaphors are not loved for themselves. Good metaphors are embraced because they do something. They are not just a pretty face. If, as a scientist, you take up a metaphor, you want some action, some benefit from it. To refer to someone (mentioned earlier) immersed in his regular scientific activity, Edward O. Wilson used the metaphor of a division of labor when looking at the leaf-cutter ants, the *Atta*. Why? Because the metaphor raised questions that yielded answers. Why are there different castes of ants, and what does each do? Could one caste do the work of another? What are the ratios of members of one caste to members of another caste, and to what extent must we factor in size and cost of production? If producing a cutter takes half the energy of producing a soldier, does this alone account for the differing numbers? Is it always more efficient to produce different forms, or can one form be used for different tasks at different stages of its life?

The same is true of the mind/brain-as-a-computer metaphor. Psychologists have taken it up because they can do something with it. And in the last half-century they have done very much with it. For instance, apart from very important advances on issues like vision, there has been a huge amount of work on the patterns and forms of human thinking, cognition. Most obviously, we might start with logically correct modes of thought. Perhaps the secret to reasoning lies in such formally valid inferences as *modus ponens* and *modus tollens*. (Affirming the antecedent and denying the consequent of a hypothetical statement.) But, as we saw in the last section, humans do not always (some might say, not often) do and think what the logicians and mathematicians prescribe. At least, they do not explicitly and formally build and follow tight deductive proofs. So, computer-inspired alternatives have been suggested. Perhaps we follow certain rules or strategies. Quick and dirty solutions, as one might say. Perhaps we make short-cut inferences that can go wrong, but generally are pretty reliable. In support of this hypothesis, consider the story of Big Blue, the chess machine that beat the world champion Gary Kasparov. At first, the machine simply used brute power to compute millions of possible moves. But it had to rise above this, because too quickly the millions multiplied beyond count. So it had to follow (have built in) various strategies of promising play given certain board configurations – which of course is precisely how humans play chess. Think ahead and look for good opportunities that you hope your opponent is missing. But recognize that here, as elsewhere in life, your strategy is not foolproof, and although it probably will work, it may not. Using the computer model as a guide, cognitive scientists can now go on to try to elucidate and classify the thinking strategies that humans actually do use, and to discover why they usually succeed and why sometimes they don't.

There are other approaches and problems that engage cognitive scientists. All to the end of what the Canadian philosopher Paul Thagard calls the central hypothesis of cognitive science, CRUM, or the Computational-Representational Understanding of Mind. "Thinking can best be understood in terms of representational structures in the mind [pictures, symbols, and so forth] and computational procedures that operate on those structures" (Thagard 2005, 10). Of course, the computer model can go only so far. We are, after all, made of meat. We do not carry around an IBM ThinkPad beneath the skull. But this does not stop cognitive scientists from trying to bring the metaphor to ever-greater fulfillment. Even twenty years ago – a long time in the history of cognitive science – we find the Princeton psychologist Philip Johnson-Laird trying to explain how computer modeling is getting ever more exact in mapping human memory. He worried about the differences: "Computers have a main memory (RAM) that stands midway between permanent long–term storage (on tape or disk) and the fleeting short-term record of the intermediate results of computations (in registers). People, however, appear to be equipped with a long-term memory that is both easy to access (RAM) yet relatively permanent (like tape or disk). They also have sensory memories, whereas computers, which for the most part lack sense organs, have no such memories" (Johnson-Laird 1988, 151). But then he comforted himself, because even then human, short-term memory was being broken down into a number of components that could be identified with the parts and processes of the computer.

Something known as "connectionism" is a major topic of interest here. The human brain functions as it does thanks to networks of special cells called "neurons." These receive information from other neurons and in turn pass it on to yet more neurons (this process is known as "synaptic transmission" and can involve both chemical and electrical elements). As the information is passed through the network, it is massaged and transformed, from raw data to conclusions. By example, a particular neuron might receive information from two other neurons. According to the strength of the inputs, this neuron will then pass on its information to other neurons down the net, and as they receive their information they in turn will pass on to others. Artificial neural networks simulate the real-life networks, and (although obviously crude compared to real life) have already shown themselves capable of remarkable feats. For instance, they can learn irregular verbs and form the correct new ones, hitherto not encountered. (Comfortingly, they make the same mistakes as do human children when learning languages – "breaked" rather than "broke," for instance.)

No one is pretending that the artificial exactly models the natural. Thagard argues that artificial networks "are similar to brain structure in that they have simple elements that excite and inhibit each other." Nevertheless, he cautions: "But real neural networks are much more complicated, with billions of neurons and trillions of connections. Moreover, real neurons are much more complex than the units in artificial networks, which merely pass activation to each other. Neurons have dozens of neurotransmitters that provide chemical links between them, so the brain must be considered in chemical as well as electrical terms. Real neurons undergo changes in synaptic and nonsynaptic properties that go beyond what is modeled in artificial neural networks" (Thagard 2005, 127). The point is that this is an ongoing research program that justifies itself both by the theoretical insights it yields and by the practical applications that it promises. We have come a long way from Descartes and his decree that the human mind is something that lies outside of and beyond the material. It is all computers now.

Whether we have come far enough to satisfy Descartes – whether we have come as far as we can or we should – is another matter. It is one that we shall raise in subsequent discussion. For now, with the science of thought and action firmly committed to the metaphor of the computer – the ultimate machine of our age – let us bring the historical survey to an end. Modern science – science at least since the sixteenth century – is dominated by the idea of the world and its parts as a machine, or machines. That is a fact that now is surely beyond question. We can therefore start to move our attention from history to analysis. This we shall do in the next and succeeding chapters.

FIVE

———

UNASKED QUESTIONS, UNSOLVED PROBLEMS

I have not been writing a general history of Western science since the Greeks. Rather, I have been trying to show the success and triumph of the machine metaphor since the Scientific Revolution in the sixteenth and seventeenth centuries. This is the root metaphor that replaced the earlier root metaphor of the organism. Admittedly, even with this more limited aim, you might question the treatment I have given. What about the gaps? Most obviously, the twentieth century saw many exciting and unexpected advances in the physical as well as other areas of science. Is it enough, for example, to have left physics back at the time of Newton? Should I not have looked at subjects like relativity theory and quantum mechanics? And the answer, of course, is that I might well have looked at such subjects and a great many more, but my suspicion – to be stated and assumed if not really argued for – is that the overall conclusion would not be changed at all. The machine metaphor triumphs.

Certainly there are areas of modern science where this conclusion is indisputable. Take the revolution in geology, with the claims about continental drift powered by plate tectonics (Ruse 1981). Here we have blind laws grinding eternally – the large plates appear from under the sea, move slowly across the surface of the globe carrying the continents on their backs, and then are reabsorbed as they disappear back into the bowels of the earth. A more machinelike picture it would be hard to imagine. And certainly the thinking about what makes for a machine has changed dramatically over the years. Machines are no longer pulleys and wheels and gears, powered by water or by heat or by animal or human brute force. The discovery of electromagnetic forces in the nineteenth century by Clerk Maxwell showed that the idea of a world of simple masses, atoms, is no longer tenable. Whatever may be the basic stuff of existence, forces in some sense must be included. For this reason, a lot of people (myself included) hesitate to speak of themselves as materialists, if this means that some kind of Cartesian *res extensa* is the substance of reality. Forces are part of this reality. However, the discovery of such forces simply reinforced the machine metaphor. After all, it is a rare complex machine today that does not make use of electricity in some form or another. Certainly again there has been the introduction of statistical thinking into science, and if quantum mechanics is right, then this kind of thinking is not simply a confession of ignorance but something that speaks of the true nature of the world. But again, as the very name quantum *mechanics* suggests, there is little or no reason to think that the machine metaphor is now rejected. There are, after all, machines that make use of the strange properties of atomic particles. The atomic clock, which depends on changes in the energy states of atoms (celsium-133 for the International System of Units), is the most obvious example.

I am going to conclude, therefore, that my overall point is now established. The machine metaphor rules modern science. I do not claim that everyone is happy with this. Indeed, in the next chapter I will consider the thinking of those who want to think outside the metaphor. But the triumph of the metaphor is the taking-off point for the next part of the case I want to make. Indeed, it is this very triumph that does lead us forward. Remember the original discussion about metaphors. Strictly speaking, they are false. They are not literal descriptions of reality. They are ways of looking at reality that give or yield great insights. To speak of a colleague as a snake in the grass is not to say that he has no limbs and slithers around the office floor on his belly. It is to say that he is sly and untrustworthy and dangerous, all predicates that our society ascribes to snakes – owing partly to biology (they are dangerous) and partly to culture (remember Genesis). Likewise, to say of another colleague that she is a real gem is not to say that she is hard and glittery or expensive, but that she is someone of great worth who is prized for what she is, and that she will not change features overnight. Notice also that, because metaphors are not literal, there are certain features of reality that they cannot explain – that they do not even set out to explain. I am simply not talking about these features. I am not talking about the mathematical abilities of either of my colleagues, nor (usually) am I talking about their religious affiliations.

This is true of the metaphors of science, a point emphasized by Thomas Kuhn in his thinking about science. His basic unit of integration, the paradigm, is something that sets the stage for doing science – it defines and suggests the problems and also, as Kuhn stressed, it rules out certain questions and lines of inquiry, as not worth pursuing. Again and again, Kuhn identified paradigm thinking with metaphorical thinking – in basic respects, a paradigm is a metaphor, and this is precisely why it has such great heuristic strength (Kuhn 1977, 1993). It lights the way but at the same time restricts the view. One's attention is focused on one set of problems and not others. And one can certainly see how this is true of the more specialized metaphors of science. For instance, when Edward O. Wilson was working on his ants, using the metaphor of a division of labor, he was focused on the ways in which they use their time and energy efficiently. This is what the division of labor is all about. He was not thinking about other problems, some of which might be genuinely interesting and important – for instance, whether the ants evolved separately from the bees and wasps or whether they were once all one line – and some of which might probably be totally spurious and certainly not worth considering – for instance, whether the worker ants feel a religious reverence toward the queen.

This restricting of the questions, this putting on the blinders, is also true of the big metaphors, the root metaphors, including the machine metaphor. We have looked in detail at the triumphs. Now raise the other side, not so much the failures but the areas where the metaphor does not go and where the scientist therefore is not led. I want to argue that there is a set of problems that are genuine, but that are not touched by the metaphor. On them, the metaphor is silent. You will learn that I do not think that membership in this set is given from on high. There are debates about what should or should not be in the set, and not everyone will agree with my choices. This is an important point, although settling it is not crucial to my argument.

In fact, what is crucial – or at least very interesting – is that the set is open-ended. There are also debates about whether or not all of the problems are genuine. I shall address this concern; but I will state right now that I do not take the inability of the machine metaphor to answer them to be in itself a reason to think them spurious. I do stress that, for the moment, I am not directly concerned with the answers to the problems, although this will certainly be an issue to be tackled later. Please do not anticipate or prejudice my argument by thinking that I am trying to trap you into agreeing to things now only to reveal later that I am pushing a religious agenda.

ORIGINS

The first of my unanswered questions is what the well-known philosopher of physics Adolf Grünbaum (2007) refers to as the *Primordial Existential Question*: Why is there something rather than nothing? Let me say straight out that I find this to be a deeply meaningful question. In the words of the philosopher J. J. C. Smart: "That anything should exist at all does seem to me a matter for the deepest awe." However, there are now two subsidiary questions. Is this primary question one that is unanswered by science? Is this a genuine question despite my personal feelings? As Smart goes on to say immediately: "But whether other people feel this sort of awe, and whether they or I ought to is another question" (Smart 1955, 46).

Turning to the first subsidiary question, I will simply say that I do not think that it is tackled by science at all and that, given our understanding of the machine metaphor, we can at once see why. I don't think that these are particularly contentious issues. We can certainly take the causal chain back to the Big Bang. As an evolutionist, that is what I spend my life doing. For myself, I am hesitant to stop there. I don't really see why there should not be something earlier, and if not something earlier with respect to our universe, then why there should not be other universes that have appeared and now exist no longer – or that are older or younger or yet unborn or whatever. It seems to me a bit ontologically over the edge to say that we and we alone could exist. At least, it would seem to require a kind of existence argument in reverse, showing that some apparently possible physical things are in fact impossible. And making that kind of argument seems to have exactly the same problems as proving that there are necessary physical things. (More on this in a moment. To block one obvious objection, I should say that I do not consider round-square cupolas as even apparently possible. And note also, I am here talking about *physical* things.)

The crucial point is that, however far back we go, an ongoing chain does not explain why there is something rather than nothing. Saint Thomas Aquinas recognized this. More recently, it is a major theme in *The God Delusion* by Richard Dawkins. You need something to break the succession. Dawkins does not think this is possible. Aquinas was more optimistic, but he realized that – a possibility Dawkins does not consider – the way to break the succession is by pointing out that we are not looking for an end point to the chain, way back in the past, even if there is one. To speak of a First Cause is not to speak in a temporal fashion at all, even though it may just so happen that there was a point at which the chain did start. We have to be looking for a cause that makes the whole thing happen, then, now, and in the future (Hick 1961). If you like to think of time as horizontal, then we are looking for a vertical cause. Something that keeps the whole kit and caboodle in being and in action. And it is pretty clear – at least this is the traditional answer, and I see no reason to give it up – that this is going to require a being (let us call it that without prejudice) that is sufficient unto itself, needs no cause, because in some sense it is necessary. In a sense, it is outside time. You do not ask when $2 + 2 = 4$ became true or when it will cease to be true. That is just not a sensible question. And the same has to be true of this being. And hence, obviously, this being is not going to be one of the chaps, a contingent fellow along with the rest of us. It cannot be a physical being, in the sense that we understand physical. That keeps us trapped in the chain. It has to be transcendent, whatever that might be.

Notice, I am not saying that any of this is true. I am simply spelling things out. And having done so, I can conclude that whatever else this being may be, he/she/it is not a subject for science. Almost by definition. It is the thing that (supposedly) keeps science going. It stands behind reality as we know it. Moreover, our history gives us a very good reason why it is not a subject for science. Machine talk puts you in mind of Hannah Glasse's recipe for jugged hare. "First catch your hare." Machines are made out of things that are supplied. You build an automobile from steel and chromium and rubber and so forth. In discussing the working of the machine, you may well talk about the materials, for instance, about putting a soft metal with a hard metal to reduce friction. But although where the metals came from may be of interest to the manufacturer, to the engineer and the user such inquiry is irrelevant. Machine talk is simply not about the question of origins.

A MEANINGLESS QUESTION?

Grant, then, that the primordial question is not one of science; the big question is whether it is a genuine question at all. We may think that it is a genuine question – I can assure you that most of my life I have thought it a genuine question – but perhaps we are mistaken. I have certainly been mistaken about some very important things that I have faced or believed or acted upon. It was Leibniz (1714b) who famously asked the question. "*Why is there something rather than nothing? For 'nothing' is simpler and easier than 'something.'*" Heidegger (1959) agreed, calling it the fundamental question of metaphysics. Yet about fifty years ago, under the influence of Wittgenstein, the popular answer was that it is a meaningless question, a question that can have no solution and hence is not genuine. As he said in the *Tractatus*, if a question can be put at all, "then it *can* be answered.... [D]oubt can only exist where there is a question; a question only where there is an answer, and this only where something *can* be *said*" (Wittgenstein 1923, 6.5, 6.51). Spelling things out in more detail, in a lecture apparently given either in 1929 or 1930:

> If I say "I wonder at the existence of the world" I am misusing language. Let me explain this: It has a perfectly good and clear sense to say that I wonder at something being the case, we all understand what it means to say that I wonder at the size of a dog which is bigger than anyone I have ever seen before or at any thing which, in the common sense of the word, is extraordinary. In every such case I wonder at something being the case which I *could* conceive *not* to be the case. I wonder at the size of this dog because I could conceive of a dog of another, namely the ordinary size, at which I should not wonder. To say "I wonder at such and such being the case" has only sense if I can imagine it not to be the case. In this sense one can wonder at, say, the existence of a house when one sees it and has not visited it for a long time and has imagined it had been pulled down in the meantime. But it is nonsense to say that I wonder at the existence of the world, because I cannot imagine it not existing. I could of course wonder at the

world around me being as it is. If for instance I had this experience while looking into the blue sky, I could wonder at the sky being blue as opposed to the case when it's clouded. But that's not what I mean. I am wondering at the sky being *whatever it is*. One might be tempted to say that what I am wondering at is a tautology, namely at the sky being blue or not blue. But then it's just nonsense to say one is wondering at a tautology. (Wittgenstein 1965, 8–9)

The trouble with this sort of thing is that people have rival intuitions on these matters. J. J. C. Smart – and Michael Ruse, for that matter – do seem to be able to imagine there being nothing at all. That is what makes the whole issue so fascinating and, as Smart says, somewhat awe-inspiring. Turn then to Paul Edwards, the editor of the *Encyclopedia of Philosophy*, who pushed the Wittgensteinian line of argument, going a little further. Apparently there is no answer that will satisfy. Suppose you invoke God, supposing him to be a necessary being, in the sense just discussed. Your question does not stop here. It seems that you need an answer beyond this, and no such answer can be given. "Voltaire, who was a firm and sincere believer in God and who never tired of denouncing atheists as blind and foolish, nevertheless asked, at the end of the article 'Why?' in his Philosophical Dictionary, 'Why is there anything?,' without for a moment suggesting that an appeal to God's creation would be a solution" (Edwards 1967, 300).

What if you are a little less demanding than Voltaire and would be satisfied with an answer in terms of a necessary being (not necessarily identified with the Christian God)? In other words, you are going part way with Wittgenstein. You disagree about the world of physical existence. You can imagine what it might be like not to have this kind of existence. But you agree that it is hard to make sense of questions about why a necessary being should exist. Necessary beings exist out of necessity, and this is tautological or pretty close. This, of course, is understanding that the possibility of a necessary being is necessary for a creator but not necessarily sufficient. (More on this in a later chapter.) You might say that you are not quite sure what a necessary being really is, but you do know that a necessary being would speak to the existence of the contingent world in which we all live. After all, you might say that we can ask meaningful questions despite not knowing the answers. It is enough that the answers make sense, that the answers are plausible. When you first ask it, you might not have much idea about the answer to the questions, "What is the meaning of life?" But if later you fall in love for the first time, then you might truly respond that now you know the meaning of life. The same is true of necessary beings. You don't know what they are, but you'll know one when you see one. For people like Edwards, however, the ultimate question seems not to fall into that category, because there is no possible answer that would satisfy. "It is not objectionable to condemn a question as meaningless on the ground that the questioner does not know what he is looking for if in the context this is a way of saying that he has ruled out all answers *a priori*; and very probably those who express themselves in this way do not mean to point to some *contingent* incapacity on the part of the questioner but, rather, to a disability consequent upon the logical possibility of obtaining an answer to the question" (Edwards 1967, 301). What is at stake here, obviously, is the claim that the notion of a necessary being is incoherent. You may not know the meaning of life, but you can make sense of the concept in some way. This is not possible for a necessary being. Hence, there is no way in which one can answer the fundamental question, because the only satisfactory answer depends on the existence of a necessary being. Therefore, there is no genuine question.

MATHEMATICAL PLATONISM

I certainly will agree that the notion of a necessary being is not an obvious one, and I have agreed already with the empiricists like David Hume that the objects of this world are certainly not candidates for the job. One can imagine any of them as not existing, and hence they cannot be necessary. "Whatever *is* may *not be*. No negation of a fact can involve a contradiction. The non-existence of any being, without exception, is as clear and distinct an idea as its existence. The proposition, which affirms it not to be, however false, is no less conceivable and intelligible, than that which affirms it to be" (Hume 1777, section xii, part 3, para. 132). But we know all of this already. The question is not whether there are necessary beings, one at least – Edwards does not demand that – but whether the search for them is remotely plausible. Or is finding them ruled out on a priori grounds before we start? I confess that I am not much given to these metaphysical speculations, but – based on vivid memories of an undergraduate major in the subject – I think a positive case might be made by analogy with mathematics. Bertrand Russell used to joke, and he should have known, that all mathematicians are

Platonists. They think they are discovering truths about real things rather than making it all up. I am not sure that this is true, but it doesn't seem to me to be a stupid position to take. Consider two discoveries. The first is in biology and concerns the homologies revealed by the study of DNA. We now know that the human and the fruit fly share almost identical stretches of DNA and that these stretches do the same things in humans and insects – they code for the order of development (Carroll et al. 2001). This is an astounding discovery – the late evolutionist Ernst Mayr (1963) denied absolutely that such parallels would be found. And it is a discovery. It is not an invention of the human mind. Now take mathematics and the Euler identity: $e^{\pi i} + 1 = 0$. Is this a discovery or a creation? It certainly does not seem like a creation. What about the Leibniz formula? $\pi/4 = 1 - 1/3 + 1/5 - 1/7 + 1/9 \ldots$?

Students of philosophy know full well that in the twentieth century – ever since a number of thinkers (the "logicists") tried to derive mathematics from logic – the putative existence of mathematical objects has been one of the most-discussed topics in philosophy. The leading logicist Gottlob Frege was an out-and-out Platonist (Hale and Wright 2001). Against this, others have been no less committed to some form of nominalism, denying that there are any abstract entities, including mathematical entities (Burgess and Rosen 1997). In their opinion, to make assumptions about existence on the basis of the form of mathematical claims – "there exist three prime numbers between 10 and 20" – is to fall into a linguistic trap. To argue otherwise is on a par with thinking that the sentence "No one runs faster than John" implies that John is at most the second-faster runner, and that some person called "no one" came in first. The philosopher of science Rudolf Carnap (1956) belonged to this camp, arguing that (sound familiar?!) it is meaningless to try to tie down mathematics with existence in any sense that we can understand. In a like vein, Paul Bernacerraf (1973) argued (in what is known as the "epistemological objection," because it depends on conditions for knowledge) that the trouble with Platonism is that it is hard to see how the abstract entities of mathematics could affect and thus get into explanations of the doings of the concrete entities of physical existence. If I think (correctly) that there is a dog in front of me rather than a cat, in some sense the dog is part of the causal nexus that leads to my belief. How could an abstract entity (or entities) possibly be part of the causal nexus that leads to my belief about the Euler identity?

Responding in favor of Platonism, one can point out that members of Carnap's generation were committed to a somewhat extreme form of empiricism. There are those who see this as the soft underbelly of the anti-Platonist thinking. We don't directly experience many of the entities of science – genes, molecules, and so forth – and since mathematics seems no less necessary for science than do these other entities, perhaps on these grounds we can credit mathematics with existence. Thus W. V. O. Quine, the midcentury doyen of American philosophers:

> Ordinary interpreted scientific discourse is as irredeemably committed to abstract objects – to nations, species, functions, numbers, sets – as it is to apples and other bodies. All these things figure as values of variables in our overall system of the world. The numbers and functions contribute just as genuinely to physical theory as do hypothetical particles. (Quine 1981, 149–50)

Kurt Gödel, responsible for one of the most celebrated and important mathematical discoveries of the twentieth century (about the limits on finding proofs of all true mathematical claims), was explicit in thinking that we have a special kind of mental organ for intuiting mathematical truths. He spoke of the "Platonistic view" as being "the only one tenable," clarifying that thereby "I mean the view that mathematics describes a non-sensual reality, which exists independently both of the acts and the dispositions of the human mind and is only perceived, and probably perceived very incompletely, by the human mind" (Gödel 1995, 3, 322–3). Admittedly, this kind of thinking is not very popular in philosophical circles, but – apart from the puzzle about how else you explain how people of completely different cultures and backgrounds can come to the same discoveries – when you start to look at great mathematicians and their modes of working, the intuition hypothesis doesn't seem quite as implausible. The great Indian mathematician Srinivasa Ramanujan simply had a sense leading him to totally unproven conjectures that later proved both true and very, very significant. His mentor, the English mathematician G. H. Hardy, remarked of the work that the theorems "must be true, because, if they were not true, no one would have the imagination to invent them" (Kanigel 1991, 168; see also Steiner 1975.)

Obviously one could spend all day going back and forth on this debate. Let us cut off and move on. Charles Dickens, in writing *David*

Copperfield, could have had David's first wife linger on, preventing the possibility of a second marriage. Dickens did not write the novel that way, but that was his choice. In *Great Expectations* he could not make up his mind about whether Pip was to marry Estelle, and so he left two endings. The Euler identity is not up for grabs like that. You cannot switch around plus and minus signs as you will. The same is true of the Leibniz formula and thousands of other equations. I am not saying that mathematical objects do exist, and I recognize that there are other philosophies of mathematics that deny reality to the objects of calculation. But I am saying that I don't think it stupid to think that the objects of mathematics do exist – π is a real entity, with properties or abilities to enter relationships of the kind just mentioned. Obviously, if it does exist, it exists in some dimension that is not of this world, and just as obviously, it exists eternally. I don't find this beyond the realm of possibility in Edwards's sense, and hence, since it seems to me that we are here talking in some sense about necessary existence, the question about necessary existence can be meaningful. And this means that the ultimate question can have meaning. Whether or not a necessary being responsible for this world could be like the truths of mathematics is perhaps another question. But I am not sure that it is our question at this moment.

AGAINST LEIBNIZ

Let's have a look at another attempt to knock down the fundamental question. This is by Adolf Grünbaum himself. He wants to argue that it is a pseudo-problem (Grünbaum 2007). He points out (truly and usefully) that the primordial question has not seemed pressing in every culture or to every inquiring mind. The ancient Greeks, for instance, never thought of it. It is very much a Judeo-Christian, and primarily Christian, issue, given the Christian doctrine that God created from nothing – *creatio ex nihilo*. Of course, this does not, as such, make the question unmeaningful – there are lots of questions that we worry about today that the ancient Greeks did not worry about – but Grünbaum uses the history to soften us up to the idea that what we think of as obvious (in the sense of obviously important and meaningful) might just be an historical artifact of the religious tradition within which we live. And if this is so, then Grünbaum invites us to ask the question: Why should we think it mysterious or significant that there is something rather than nothing? We only ask the primordial question because we do think it mysterious or significant. The answer Grünbaum gives to his question (and I have no reason to doubt this) is that it is because of Leibniz's feeling or claim that nothing would be simpler than something, and hence an empty world would be more likely than a full one like that in which we live – "empty" here meaning "the putative state of affairs in which no contingent objects exist at all" (Grünbaum 2007, 444).

It is here that Grünbaum registers his objections. First, on a priori grounds, he can see no reason why the simplest should obtain, be the case. He faults Leibniz for not spelling out the sense in which simpler is more probable – remember, Leibniz talks of nothing being "easier" than something, and Grünbaum objects that this is meaningless talk. Second, on a posteriori grounds, on empirical grounds, he does not see why empty (in the sense just specified) should be the case. He draws our attention to two cosmological hypotheses about the world, one being that which sees a spatially closed universe that expands from the Big Bang, reaches an outer limit, and then contracts again. (This is the model of the Russian mathematician Alexander Friedmann.) The other is the (now admittedly discredited) steady-state universe in which matter appears spontaneously and constantly. This "steady-state world features the accretion or formation of new matter as its natural, normal, spontaneous behavior. And although this accretive formation is indeed out of nothing, it is clearly not 'creation' by an external agent" (Grünbaum 2007, 449).

Neither of these arguments seems overwhelming. One can make some sense of what is meant by "less is more likely." Suppose, to take an example of Erik Carlson and Erik J. Olsson (2001), one has two identical universes, both having a coin-tossing machine giving entirely random results. The machines are set in motion, and a penny is tossed, action ceasing when the coin comes up tails. The one universe tosses the penny once, and tails comes up. The other universe tosses the penny a billion times before the penny comes up tails. Surely in some sense the first universe is simpler, easier than the second, and much more likely? At another, more intuitive level, surely we do feel surprise when there are things, unless we have been given some explanation? I am out walking, and I come across an absolutely spotless old automobile in a field – I do expect some explanation. If I am out walking and do not come across anything, I don't find myself saying: "I wonder why there is no mint edition

Model T in this field." Note that I am not saying that if I find something I expect intelligence. That was the assumption of Archdeacon Paley on finding a watch in the wilderness. I could well accept a proximate explanation in terms of laws. But I do expect an explanation, because the presumption is that one will have nothing.

This is obviously an experientially based presumption, but I don't think anyone doubts that the primordial question starts with experience. The question is whether experience can answer it, and my suspicion is that it cannot. This brings in Grünbaum's second point, because he seems to think that experience (or at least an empirically based view of the universe) can be definitive. But especially in the light of what has been said already, I am not sure that this is so. On either cosmological scenario, it still seems to be open for someone at the empirical level to ask why these things came into being. Perhaps not within the context of the theories themselves, but around and supplementing the theories. More than this, surely today answers would be proposed. Especially in the steady-state case, the presumption would be that something was going on. As the philosopher David Lewis (2004) rightly remarks: "The more we learn about the vacuum, the more we find out that it is full of causally active objects: force fields, photons, and "virtual" particles. Space-time itself, if curved, can serve as a repository of energy. And perhaps that is not the end" (p. 277). But even if there were no answers forthcoming, the objection would be that the primordial question is not about the empirical facts of the matter or their causes in this sense. It is a metaphysical question about why the whole system holds in place. At a certain level, Grünbaum ignores this as he argues that the question is based on a pseudo-problem.

I conclude, therefore, that the primordial question is not answered by science and is not obviously a meaningless question or directed toward a pseudo-problem. There are those, in the tradition of Leibniz, who try to answer it (for instance, Rundle 2004; Van Inwagen 1996; Parfit 1998a, b). This is not my concern here. Later, however, I shall ask how appropriate it is for Christianity to speak to the question.

MORALITY

Machines can be used for good or ill, but they are not in themselves moral objects. It is how they are used and why that makes for morality. For instance, I am deeply opposed to capital punishment, and hence I can think of little good use for the electric chair. I suspect that there are many who, if faced with the option of putting Adolf Hitler into the chair and pulling the switch, would do so without hesitation and feel a strong moral glow afterward. This being so, we should not expect something based on the machine metaphor to be a source of moral judgments. The world, inasmuch as it is – or rather, inasmuch as it is seen to be – a machine, is not something that tells you what to do, that distinguishes between right and wrong. Of course, technology, like training and ability, can influence moral decisions. We can now properly attempt open-heart surgery in a way closed to the Victorians. But the morality of trying to help someone with heart disease is unchanged.

The point I am making is, of course, the point made most clearly by Hume. The way that things are, the objects of the world, cannot tell us about the way that things should be, the dictates of morality.

> In every system of morality, which I have hitherto met with, I have always remarked, that the author proceeds for some time in the ordinary way of reasoning, and establishes the being of a God, or makes observations concerning human affairs; when of a sudden I am surprized to find, that instead of the usual copulations of propositions, is, and is not, I meet with no proposition that is not connected with an ought, or an ought not. This change is impercep-tible; but is, however, of the last consequence. For as this ought, or ought not, expresses some new relation or affirmation, it is neces-sary that it should be observed and explained; and at the same time that a reason should be given, for what seems altogether inconceiv-able, how this new relation can be a deduction from others, which are entirely different from it. But as authors do not commonly use this precaution, I shall presume to recommend it to the readers; and am persuaded, that this small attention would subvert all the vulgar systems of morality, and let us see, that the distinction of vice and virtue is not founded merely on the relations of objects, nor is perceived by reason. (Hume 1739, 302)

So here we have another set of questions that science does not even set out to answer. It is important, however, to clarify which questions science does not set out to answer, and which questions (I believe) science can answer. The claim is certainly not that science, even discounting the matter of technology made earlier, can say noth-ing about morality. I happen to believe that it can say a great deal about what we do and think and in particular about why we think some things are moral and some things are not (Ruse 1986, 2006). As we saw in the last chapter, building on the insights of Charles Darwin in his *Descent of Man*, and particularly drawing on the work of others like comparative anthropologists, today's evolutionary biologists – especially those interested in the origins and nature of human behavior, the human sociobiologists (or as they like to call themselves today, the evolutionary psychologists) – have done much to show how and why it is that humans act morally. Contrary to popular belief, we humans are not naked apes, with blood-drenched fangs, unable to restrain our brute natures as we rape and pillage our way through history. The struggle for existence – more precisely, the struggle for reproduction – takes many forms, and not all are violent. Humans, for better or worse, are highly social animals, and we need adaptations to make our sociality a functioning phenomenon. Morality is high on the list. For obvious reasons, we are selfish in the sense of self-regarding and self-concerned – if we were not, we would never get a square meal, let alone a mate. But we need to get along, and simple utilitarian calculation – you scratch my back and I'll scratch yours – is never going to be enough. We need some force, emotions, to make us work with others to make us give spontaneously, as it were. This is where morality comes in.

Note that morality is not simply a recipe for suckers. Part of morality is enforcing the behavior of others, ensuring that they do their bit as we do ours. For this reason, one might with reason say that the supreme principle of morality is that of being fair. It is certainly this which rules the nursery – "It's not fair, Johnny has the biggest half" – and according to John Rawls (1971), the leading American moral philosopher of the past half-century, it is this which is the basis for all moral action. "Justice as fairness." This is a form of social contract theory. How would we want society constituted – pay and medical care and so forth – if we did not know what role would be allotted to us (if we were "behind the veil of ignorance")? We could be female, born of rich parents, healthy, and beautiful; or we could be male, born of poor parents, sick, and ugly. Rawls argues that we want society set up so that whatever place we find ourselves in, we would benefit the most given the risks. We cannot just go for the female role, because we might end up with the male role. Hence, we want a society that will look after the male as well as possible. This does not necessarily mean that everyone will get the same. If we want good medical care then we might have to pay doctors twice the amount we pay professors. Rather, we want a society where the loser in birth's gamble gets as good a deal as possible.

Rawls fully admits that this is all talk about hypotheticals. No one thinks that societies were set up by a gang of leaders and then the rules made mandatory. However, perhaps our genes did what our ancestors did not.

> In arguing for the greater stability of the principles of justice I have assumed that certain psychological laws are true, or approximately so. I shall not pursue the question of stability beyond this point. We may note however that one may ask how it is that human beings have acquired a nature described by these psychological principles. The theory of evolution would suggest that it is the outcome of natural selection; the capacity for a sense of justice and the moral feelings is an adaptation of mankind to its place in nature. As ethologists maintain, the behavior patterns of a species, and the psychological mechanisms of their acquisition, are just as much its characteristics as are the distinctive features of its bodily structures; and these patterns of behavior have an evolution exactly as organs and bones do. It seems clear that for members of a species which live in stable social groups, the ability to comply with fair cooperative arrangements and to develop the sentiments necessary to support them is highly advantageous, especially when individuals have a long life and are dependent on one another. These conditions guarantee innumerable occasions when mutual justice consistently adhered to is beneficial to all parties. (Rawls 1971, 502–3) [In support of his position, Rawls footnotes human sociobiologists.]

Actually, as the evolutionary psychologists are telling us, things are a bit more complex and interesting than this. It certainly seems that our moral sentiments are shaped in even more distinctive ways by our evolutionary past. The trolley problem, discussed in the previous chapter, shows this clearly. Another example comes from the fact that, despite exhortations by some moralists to love every other human being indifferently, we do not, nor do we often think that we should. "Charity begins at home." I like to refer to this as the *Bleak House* problem or issue. There, Dickens is scathing about those who worry about the problems of the starving poor in Africa when they are indifferent to the problems of the starving poor of England and even to the needs of their own families. There are clearly good biological reasons why we feel this way. Selection has fashioned us to think first of closest relatives, then of more distant ones and friends in our social group, and so on out. Although no evolutionist, Hume spotted this. "A man naturally loves his children better than his nephews, his nephews better than his cousins, his cousins better than strangers, where every thing else is equal. Hence arise our common measures of duty, in preferring the one to the other. Our sense of duty always follows the common and natural course of our passions" (Hume 1739, 311).

So grant that science – a machine-metaphor science – can have implications for the ways in which we think about morality, and for our actions based on such thinking. (This is the part of moral theory, ethics, known as substantive or normative ethics.) This still escapes the main point. Science cannot justify claims about the basis for our moral beliefs and actions. (This is the part of moral theory known as metaethics.) Why should I be good? Why should I be fair? Why should I care about my kids rather the kids of a stranger? Why should I care about the bloke I am about to shove on the rails? Philosophers and theologians and preachers have had lots to say on these issues. Some appeal to the will of God, some to nonempirical facts that supposedly hold eternally – a bit like mathematics, I suppose. And some have argued that there is no justification, that it is all a matter of psychology without a basis. This last position is sometimes known as "ethical skepticism," which is perhaps a little misleading because the skepticism is not at all about substantive ethics but about the metaethical foundations of morality (Mackie 1977). I believe that this is Hume's position. As it happens, I have myself endorsed a version of this kind of thinking (Ruse 1986). Truly, however, for the purposes of the discussion here, the exact truth of the matter is not relevant. What is relevant is that morality cannot be derived from the physical facts of the matter. More pertinently, it cannot be derived from machine-based science.

MINDS

Leibniz again:

> One is obliged to admit that perception and what depends upon it is inexplicable on mechanical principles, that is, by figures and motions. In imagining that there is a machine whose construction would enable it to think, to sense, and to have perception, one could conceive it enlarged while retaining the same proportions, so that one could enter into it, just like into a windmill. Supposing

> this, one should, when visiting within it, find only parts pushing
> one another, and never anything by which to explain a perception.
> Thus it is in the simple substance, and not in the composite or in
> the machine, that one must look for perception. (Leibniz 1714a,
> *Monadology*, section 17)

Machines cannot think. Hence any science that is machine-
metaphor based is bound to fail the attempt to explain consciousness – thinking, self-awareness, what Steven Pinker (1997) calls
"sentience."

There is much truth in this argument, but it needs some careful
unpacking. If brains are literally machines, then obviously machines
can think. I am not quite sure where I stand on that. I am unwilling to say that it is absolutely, logically impossible for a machine
to think. The more important question is whether looking at things
as machines helps us to explain consciousness, and there I think the
answer is a lot less certain. If, with Leibniz, you get inside a machine,
you can find out a huge amount about how it works – your mechanistic mechanism, as cognitive science shows very well, can work
wonders of explanation. But you don't touch what the psychologist
David Chalmers has called the "hard question": consciousness or
sensed experience.

> What makes the hard problem hard and almost unique is that it
> goes *beyond* problems about the performance of functions. To see
> this, note that even when we have explained the performance of
> all the cognitive and behavioral functions in the vicinity of expe-
> rience – perceptual discrimination, categorization, internal access,
> verbal report – there may still remain a further unanswered ques-
> tion: *Why is the performance of these functions accompanied by
> experience?* A simple explanation of the functions leaves this ques-
> tion open. (Chalmers 1997, 12)

Chalmers goes on to point out that if we analyze the workings of
(say) a gene, there comes a point where it is really inappropriate to
ask what a gene really is. If you have not followed the explanation
thus far – the DNA, the copying, the expression, and so forth – you
have not understood the point of the exercise. But with consciousness it is different. For: "If someone says 'I can see that you have
explained how DNA stores and transmits hereditary information
from one generation to the next, but you have not explained how it
is a *gene*,' then they are making a conceptual mistake. All it means

to be a gene is to be an entity that performs the relevant storage and transmission function. But if someone says 'I can see how you have explained how information is discriminated, integrated, and reported, but you have not explained how it is *experienced*,' they are not making a conceptual mistake. This is a nontrivial further question." Explaining, Chalmers writes:

> This further question is the key question in the problem of consciousness. Why doesn't all this information–processing go on 'in the dark', free of any inner feel? Why is it that when electromagnetic waveforms impinge on a retina and are discriminated and categorized by a visual system, this discrimination and categorization is experienced as a sensation of vivid red? We know that conscious experience *does* arise when these functions are performed, but the very fact that it arises is the central mystery. (p. 13)

There are many who agree with Chalmers on this point, even if they do not accept his solution, which is a kind of polished-up Cartesian dualism. For instance, in a modern-day version of Leibniz's argument, the philosopher John Searle (1980) invites us to consider a man locked in a room, receiving and sending out pieces of paper. On the incoming paper are symbols in Chinese, a language he does not speak, and his job is to correlate them with the similar symbols he finds in a manual, copy down the given answer, and post this new piece of paper to the people waiting outside the room. Searle points out that although this all makes sense to the people outside the room, there is no reason to think that the man in the room is any nearer to understanding Chinese. In the terminology of the linguists, he may have the syntax right – he knows how to manipulate the symbols and string them together – but he has no understanding of the semantics – how the symbols relate the thinker to the world of experience, how they give meaning. Related to this kind of argument, one can make the point against the analogy of the brain as computer hardware and thinking as computer software, that it is all very well to have the software as something etched on a disk or glowing as symbols on a screen, but unless there is now a thinking mind able to interpret these marks, they are nothing.

And that, for a lot of people, is that. In fact, on the spectrum, Chalmers is quite far to the left. Apart from the hard problem, he lists "easy problems," like how the mind integrates the information it gets, how the mind controls behavior, and how we distinguish between being awake and being asleep. In one sense, of course, they are not easy problems, but Chalmers thinks they are problems that we can tackle and have some hope of solving. Others, however, will have none of this. "There are no easy problems of consciousness" (Lowe 1997, 117). Take, for instance, the mind processing information – for example, doing the sorts of things that cognitive scientists get so excited about, like making mathematical inferences or solving logic problems. Chalmers writes: "The easy problems are easy precisely because they concern the explanation of cognitive *abilities* and *functions*. To explain a cognitive function, we need only specify a mechanism that can perform the function. The methods of cognitive science are well–suited for this sort of explanation, and so are well-suited for the easy problems of consciousness" (Chalmers 1997, 11). His critic responds: "Chalmers' problem is that he entirely begs the real question at issue in supposing that the sort of *performance* we have to do with in cases of thoughtful human activity is something that can be characterized in a mechanistic way and which, consequently, a 'mechanism' can uncontroversially be supposed capable of engaging in" (Lowe 1997, 121).

Note how the applicability of the machine metaphor is the crucial issue here. Chalmers says it can be applied to the easy questions. His critic denies this. Others make this the central point in their claims that there is no way we are going to explain consciousness given the ways we are going about it at the moment.

> The concept of the brain as machine has been favored by neuroscientists, who want a paradigm to organize our rapidly increasing knowledge of brain physiology and psychology, and by cognitive scientists as a model to guide them in their quest for machine consciousness. I think, however, that such a concept will not serve the purposes of either group. There is a vast difference between the structure of a machine, whose components have been formed with difficulty from contrary matter (which often asserts its less than ideal behavior in unexpected ways at inconvenient times), and a complex organ like the brain, whose structure has not been imposed on matter but seems to arise as an expression of the potentialities of the organic molecules themselves. A machine is constrained by design to behave in accordance with rules from without. It is hardly conceivable to me that this distinction is not relevant to understanding the relation of consciousness to the brain. (Bilodeau 1997, 230)

Expectedly, this line of argument has not gone unchallenged. The philosopher Daniel Dennett – a student, remember, of the analytic behaviorist Gilbert Ryle – is particularly scornful. He thinks that once you have located and identified the functions that the various brain parts play, that is an end to matters.

> What impresses *me* about my own consciousness, as I know it so intimately, is my delight in some features and dismay over others, my distraction and concentration, my unnamable sinking feelings of foreboding and my blithe disregard of some perceptual details, my obsessions and oversights, my ability to conjure up fantasies, my inability to hold more than a few items in consciousness at a time, my ability to be moved to tears by the vivid recollection of the death of a loved one, my inability to catch myself in the act of framing the words I sometimes say to myself, and so forth. These are all 'merely' the 'performance of functions' or the manifestation of various complex dispositions to perform functions. In the course of making an introspective catalogue of evidence, I wouldn't know what I was thinking about if I couldn't identify them for myself by these functional differentia. Subtract them away, and nothing is left beyond a weird conviction (in some people) that there is some ineffable residue of 'qualitative content' bereft of all powers to move us, delight us, remind us of anything. (Dennett 1997, 35)

Even a friendly reader might wonder if this is an argument, and, if it is an argument, how it is making its case. It seems simply to ignore the worries of people like Chalmers. I think Dennett admits that he is conscious, but then he identifies his feelings and experiences with the workings of the brain and claims that that is the solution. But the whole point is that it is not the solution. The brain firing away in certain ways is not the same as feeling lovesick or delighting at the beauty of a Haydn quartet, or appreciating the Euler identity, for that matter.

Paul and Pat Churchland at least give some arguments for their critical position. On the one hand, they are less than fully impressed by the arguments of Leibniz and Searle. Agreed, there is not much consciousness inside a mill or in the Chinese Room. But apart from the fact that we should note that if there is consciousness it is not something that would be immediately apparent to the observer or the translator – the consciousness would be of the whole system as it is of our machine-brains – they argue that the machines (assuming the

Chinese Room to be a machine with the man inside one of the parts) are way too crude to expect consciousness. But the real problem is that we are hung up on folk psychology. We think that what we believe today must be the absolute bedrock of inquiry. Our sense of consciousness must be untouched. However, they argue that that is not the way things go in science. As we discover more, as we come up with new findings and theories, so our bedrock beliefs have to be molded, changed, and sometimes discarded. Take the analogy of light.

> From the standpoint of uninformed common sense, light and its manifold sensory properties certainly seemed to be utterly different from anything so esoteric and alien as coupled electric and magnetic fields oscillating at a million billion cycles per second. And yet, the intuitive impression of vast differences notwithstanding, that is exactly what light turns out to be. Using the resources of electromagnetic theory, we can reconstruct, in a unified and revealing way, all of the intrinsic and causal properties of light, such as its traveling at 300,000 km per second, its refraction, its reflection, its polarizability, its splitting into distinct colors and so forth.
>
> In this way, visible light and a host of its nonvisible cousins (radiant heat, radio waves, gamma rays, X rays) have all been successfully identified with (i.e. reduced to) electromagnetic waves of appropriate wavelengths. Who will be so bold as to insist, just as the neuroscientific evidence is starting to pour in, that mental states cannot find a similar fate? (P. M. Churchland 1995, 206)

Completing their case, the Churchlands tell the pathetic story of Pat's science teacher back in Canada, longer ago than it is perhaps decent to mention (P. S. Churchland 1997). He was convinced that life is an irreducible concept – a happy vitalist he. But he was wrong! Who today would deny that the concept of life has been explained fully? We know about the DNA, about the cell, about physiology, and much, much more, so that mysterious entelechies and *élans vitaux* no longer glow in the innards of every living being. Molecules in motion are what we find, and molecules in motion are all we need. Could not our thinking about the mind be in exactly the same situation – pre-Cambrian, pro-Canadian, as it were? At some point in the future we shall look back and laugh (or cry) at our ignorance.

The Churchlands do make some very good points. The story of vitalism is salutary. (More on this at the end of the chapter. For now, I will agree with the Churchlands that life has been reduced

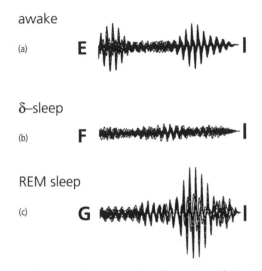

awake

(a)

δ–sleep

(b)

REM sleep

(c)

Figure 5.1. There is far more brain activity during Rapid Eye Movement sleep than during deep sleep, as is shown by these brain wave pictures. From Paul Churchland, *The Engine of Reason, the Seat of the Soul* (1995).

to molecules in motion. If you disagree, it will not affect the overall argument.) It is indeed fair to ask for realistic models of brains rather than windmills and Chinese Rooms. But, even when you have them, and even if you think you now have consciousness at some level, there is still the question of how it comes about, and why, and what is the connection with the machine? Basically, the question of what it is. When you start to get down to the details of the Churchlands' arguments, an awful lot of question begging seems to surface. Consider a discussion by Paul about sleeping and consciousness. He discusses what is going on, distinguishing waking, deep sleep, and REM sleep (Rapid Eye Movement sleep, during which we dream). He shows pictures of the different wave patterns in the brain associated with these states. As you can see, he can show how waking and REM sleep have much more brain activity than deep (nondreaming, unconscious) sleep [Figure 5.1].

All of this seems absolutely true and, as Paul points out, might lead to interesting predictions. "The account here sketched, for the nature of dream activity, may also explain why the actions and episodes in one's dreams are regularly so mundane and prototypical in character. In the absence of the usual control exerted on the recurrent system by sensory inputs, the principal determinant of the system's wandering trajectory through activation space will be the antecedent landscape of temporally structured prototypes already in place" (P. M. Churchland 1995, 222). But none of this seems to speak to Chalmers or Pinker about experience, about sentience, about the conscious awareness of thinking. We are still at the physical level – including here, of course, electricity and so forth as part of the physical. The gap remains. And this being so, perhaps (as Chalmers argues) the vitalism example does not hold. We can explain the physical in terms of the physical. The question is about explaining the mental. And while the dream work does at one level explain aspects about the mental, it does not explain why we have the mental or what the mental really is. Perhaps things will change down the road. But for now, the mental remains.

I conclude, therefore, that although it may be possible that machines could think, a machine-based science of the mind leaves the hard question untouched.

PURPOSE

The Nobel laureate in physics Steven Weinberg has said: "The more the universe seems comprehensible, the more it also seems pointless" (Weinberg 1977, 154). Likewise remember Charles Darwin: "In contrast to the old theory that all adaptation to purpose in the arrangements of the world was fore-calculated and fore-ordained, and that all organisms were merely wheels in a gigantic machine made once for all, and incapable of improvement, this new view is so grand that it deserved a higher appreciation than it has ever met with" (Darwin 1879, 224). Richard Dawkins again:

> In a universe of blind physical forces and genetic replication, some people are going to get hurt, other people are going to get lucky, and you won't find any rhyme or reason in it, nor any justice. The universe we observe has precisely the properties we should expect if there is, at bottom, no design, no purpose, no evil and no good, nothing but blind, pitiless indifference. As that unhappy poet A. E. Houseman put it:
>
> > For Nature, heartless, witless Nature
> > Will neither know nor care.
>
> DNA neither knows nor cares. DNA just is. And we dance to its music. (Dawkins 1995, 133)

What is going on here? Machines are artifacts, and they have purposes. You make a windmill to grind your corn. You make an automobile to carry you from one place to another. You make a refrigerator to keep your food cool so it will not spoil. Yet here are scientists telling us that the world has no purpose and, in Darwin's case, congratulating themselves on having dropped the machine metaphor. It seems that we have something of an inversion here. Machine metaphors imply purposes. Scientists do not want purposes. Hence the machine metaphor cannot apply.

Let us work backward on this, from cognitive science, through biology, to physics. Of one thing we can be sure. The machine metaphor is alive and well in the science of the brain. The brain is a computer – metaphorically a laptop like the one I am typing on at the moment, perhaps literally a computer made of meat. That being so, since machines have purposes, the brain must have a purpose. And that purpose is not hard to find. The brain – let us stay with humans – is used to keep humans alive and functioning – reasoning, caring, loving, and everything else that brains do. The mind may not be literally software, but it is something like it, whether you think with people like Dennett that once you have described the software that is all there is to the mind, or with people like Chalmers that once you have described the software you have now got to interpret it – whether it be the information etched on the disk or the information shooting around the brain beneath the skull. Inasmuch as mind is part of the picture, therefore, I presume that we want to say that the mind has a purpose.

I doubt anyone today wants to go the Thomas Henry Huxley route and argue that we are automata, and that the mind just sits on top of the machine, doing nothing. Indeed, most evolutionary psychologists would have no problem including the mind in the machine package – however it is done, minds help the on-board computers do things that much more effectively. Thanks to minds, we can plan and decide in a way that the nonconscious, genetically determined ants cannot. If something goes wrong, they are helpless. If something goes wrong, we humans plan a course to avoid it or to repair it. Indeed, I suspect that most evolutionary psychologists would be happy to say that natural selection works on the mind just as much as it works on the brain. The two go in tandem. Whatever the brain-mind connection, once minds started into being – somewhere between bacteria and humans – selection started to work on them, and, just as better brains led to better minds, so better minds led to the selection of better brains. Although there is much debate among physical anthropologists about why human brains have grown larger over the past four to five million years, no one denies that it was a matter of going a different route from other mammals. With our large brains, we could make decisions and plan ahead and so forth, at the same time working together socially. (This is very important and, as evolutionary psychologists now understand, a very large part of our biology.) It was clearly a feedback process. Big brains need lots of protein, which in nature means large chunks of meat. We humans did not go the tiger route, becoming super-physical predators. We went the human route, learning how to work in teams and how to outfox the prey – that meant brains and intelligence, and here we now are.

So the machine metaphor is alive and well when it comes to brains and thinking. Move back now to organisms. Again we see the machine metaphor at work. I am not sure that biologists generally want to think of organisms as machines, although when Richard Dawkins speaks of organisms as "survival machines" I suppose he is saying that the purpose of organisms is to reproduce genes. Also, where we have social insects, like the hymenoptera (ants, bees, and wasps), we might say of an individual – a soldier ant, for example – that its purpose is to protect the nest. Although notice that in this latter case we are treating the nest as a supra-organism, and the soldiers as parts rather than as individuals in their own right. If one thinks of the soldiers as individuals – using kin selection, one says that the soldiers are furthering their own reproduction by raising fertile sisters – then I am not sure that one would say that the purpose of the soldier is to protect the nest. The soldier is protecting the nest, but that is not its purpose. However, when we do come to parts – unambiguous parts like eyes and teeth, as well as the parts of supra-organisms – then we do speak of purpose. Just as the purpose of the brakes is to slow down the automobile and the purpose of the wings is to lift the plane, the purpose of the eyes is to see and the purpose of the teeth is to tear and grind. Machines have functioning parts serving ends, and so do organisms. As Darwin pointed out, the machinelike nature of organisms is often a bit odd, as all sorts of things are co-opted for new purposes – the bones in the wings of birds undoubtedly started life as the bones in the fins of fish – but that of course is the result of the evolutionary process. You cannot stop life and redesign from scratch but must always cobble together from what you have at hand. But machinelike it is, whether it be the elastic from your underpants that you are using to act as substitute for your broken fan belt as you attempt the cross-Sahara car rally, or whether it be a lung once used for floatation and now for breathing.

Finally, we come back to physics and the objects of the inanimate world – pendulum to planet. The simple fact is that, as Darwin pointed out, we don't use the machine metaphor in the same way here. We don't ask about the purpose or function of the moon – except in a joking way to suggest that it is a light to guide the way home of drunken philosophers! We don't ask about purposes, because this was all part of the drive to rid science of Aristotelian final causes. We have proximate causes in science and leave matters at that – material and final causes being collapsed down into proximate causes (as they often were by Aristotle himself). So what is going on here? What price now claims about machine metaphors?

At this point we obviously must go back to our earlier discussion about the evolution of the machine metaphor, and to the two forms that it took. When the machine metaphor was introduced into science it had the simple form of a machine, with a probable purpose.

> A mechanistic world-picture is then, in particular, to be regarded as a conception according to which the physical universe is seen as a great machine, which, once it has been set in motion, by virtue of its construction performs the work for which it was called into existence. It cannot be doubted that this view was embraced by several thinkers, both in the period preliminary to classical science... and during its later development; we need only remind the reader of the comparison of the material world to an ingeniously contrived clockwork, which is frequently encountered in speculations on natural philosophy, or in Newton's idea that from time to time the Creator has to interfere in the course of the material processes in order to secure the normal progress against disturbances. (Dijksterhuis 1961, 495)

Increasingly, however, this would not do. With hardly an exception, perhaps Fermat's theorem of least distance being one, people did not find reference to final causes useful in their science – remember Bacon and vestal virgins, beautiful but useless. And so we got an evolution from machine thinking to simple mechanistic thinking. In the terminology introduced in Chapter 3, from mechanism 1 (artifact mechanism) to mechanism 2 (governed-by-laws mechanism). Not what is the part for, but how does the part work. "The mechanism of the world picture led with irresistible consistency to the conception of God as a retired engineer, and from this to His complete elimination was only a step" (Dijksterhuis 1961, 491). But if talk about God as the machine maker was no longer appropriate, it did not mean that the machine metaphor was destroyed or rejected. It had to be modified, so that only certain aspects of machines were considered, namely, their workings and not their ends.

> That the conception of the universe as a machine created by God has not performed any essential function in the development of classical science does not yet imply that the term mechanistic has nothing whatever to do with the concept of a machine. At all times there used to be a strong tendency amongst physicists, particularly in England, to form as concrete a picture as possible of the physical reality behind the phenomena, the not directly perceptible cause of that which can be perceived by the senses; they were always looking for hidden mechanisms, and in so doing supposed, without being concerned about this assumption, that these would be essentially the same kind as the simple instruments which men had used from time immemorial to relieve their work, so that a skilful mechanical engineer would be able to imitate the real course of the events taking place in the microcosm in a mechanical model on a larger scale. The pursuit of this object was, and is, frequently looked upon as the really distinctive feature of classical science and the true meaning of the descriptive adjective 'mechanistic.'
> (p. 497)

Of course we have moved on from clock mechanisms through all sorts of more sophisticated and subtle mechanisms, but the basic idea persists. The world works in ways that machines work. That is the force of the metaphor, and that is why it is no great surprise that Steven Weinberg finds no purpose. He is not looking for it. Neither is Dawkins, nor is Charles Darwin when he is talking about the bigger picture. I am not saying that if there were purposes crying out to be recognized they would not be recognized, but it is not easy to imagine what form these purposes might take, and certainly no one as a scientist is making an effort to find them.

But now this raises questions as we turn back to biology and cognitive science. If the original machine metaphor (mechanism 1) is gone from physics, can it stay on in these other subjects? At one level, the response is – why not? The world as a whole may not seem very machinelike, nor do its parts function as if they have ends, but if the organic world and the world of the brain do seem machinelike, then why not use the metaphor? More particularly, machines are things that are designed with ends in view. If organisms – their parts, that is – seem as if designed, if brain-minds seem as if designed, then why not use the full metaphor? Of course, we know that they are not designed directly. They are produced by natural selection. But that is the whole point about selection – it produces things like eyes and teeth that are "as if" designed. Eyes do have ends or purposes, namely, seeing. Teeth and brains likewise have ends or purposes (Ruse 2003).

This raises a disjunctive question. If eyes and teeth and brains have ends, are we not smuggling in Aristotelian final causes, or, if not, do we need such talk? Let us just talk in a nonteleological fashion about mechanisms, like the physicists. In reply to the first disjunct, we can certainly say we are not bringing in Aristotelian final causes. There is no question of vital forces striving toward ends. In the important sense, everything is purely mechanistic – material objects moving in space and time (mechanism 2). Eyes see and, at one level, that is that. Sometimes eyes do not realize their goal – when people have cataracts, for instance. And brains don't always work properly. Try drinking ten pints of Guinness and then doing mathematics. In reply to the second disjunct, we can use function talk (mechanism 1) because eyes and brains are as if designed and, apart from anything else, we need function talk because thinking in terms of design and ends is terrifically valuable heuristically. It leads to new discoveries. Remember the funny diamond plates down the back of the stegosaurus. What were they for? You might think that they were weapons of attack or defense, but the bone structure suggests that they were pretty fragile. Possibly they were the large breasts and massive muscles of the dinosaur world, turning on members of the opposite sex. Possibly not, because the two sexes seem indistinguishable. The popular answer today is that they were for heating and cooling the brute (Farlow et al. 1976; Ruse 2008a). Dinos were cold-blooded (don't believe the *National Enquirer*), and they needed warmth in the morning to get going and cool in midday, particularly given the heat that their fermenting vegetable diet produced. (Think of piles of lawn cuttings.) Whoever would have worked all of that out without thinking in terms of design, and realizing that those plates look just like the heat-transfer plates that you find in electrical generating stations?

In the case of brains, you might think that we have another kind of forward-looking Aristotelianism. Not only do they function, they also plan for the future, and if things go wrong, they help to put things right again. In a sense, therefore, brains – thinking, anyway – are goal-directed in the sense used by philosophers of being able to right things when they go astray. ("Goal directedness" in this sense is not just a matter of being directed to a target – that is the usual kind of teleology – but of being able to readjust in the face of barriers to the target. See Nagel 1961 and Ruse 1973.) All of this is probably true, but again we don't really have anything that would stand in the way of mechanism in the physical sense (mechanism 2). Organisms, humans particularly, aim for targets, and if things go ill they try other strategies, but there is no need to think that any of this violates mechanism at the proximate-cause level. I aim to get my Ph.D. and having failed the French examination, I take the German qualifier instead. I may succeed, but if I do not and do not get the degree, so be it. There were no vital forces striving to achieve the end, and the end was not in some sense reaching back out of the future to the present. My present motivation is the goal of getting the degree, not the actual getting of the degree. Sometimes what we aim for, what is our goal, and what is the end are very different things. Ask any person who has just been through a divorce about their hopes and goals on their wedding day.

The conclusion is that the machine metaphor lets us ask about ends when there are designlike features to the world, but when there are not, there is no place in science for end talk. Another set of questions unanswered by science.

FOUR LIMITS – MORE OR LESS?

I have talked of four areas where I claim the machine metaphor (in some sense) does not provide answers. Why is there something rather than nothing? What is the foundation of morality? What is consciousness? What is the point of it all? I do not take four – I especially do not take this four – as having some special status. If people like Paul and Patricia Churchland are right, then we are well on the way to solving the consciousness problem, and so we are down to three areas. Likewise, if the primordial question is ill conceived, there is another way to cut down the number. Conversely, you might think there are areas that should be included on the list. The vitalists would certainly include the origin and nature of life itself. Whether or not Pat Churchland's high school biology teacher is still alive, many fewer today would argue that life is some special thing that cannot be understood fully in physico-chemical terms. The cosmologist and popular science and religion writer Paul Davies (1999) is perhaps one. He apparently is searching for new laws of organization that give rise to life. But he does not specify what these would be like, so until he does come through we can rest easy. The Intelligent Design theorists are obviously others who think science inadequate at this point. Blind mechanical law apparently cannot explain organized complexity, like the motor on the bacterial flagellum (Behe 1996). Here, I am focusing on those who do at least make a moderate effort to stay within the bounds of science before they look elsewhere.

Another area you might think untouched by science is the free will problem. What is going on when we chose freely, as when you picked up this book and started to read? Along with many philosophers and I suspect most scientists, I am inclined to think that all that can be usefully said about the topic – or need be usefully said – can be said by Humean compatibilism. Free will is a matter of being beyond constraint, not of being outside the causal nexus. Clearly this is all very much bound up with the mechanical model of the world, including humans. (See Chapter 7 for a detailed account of how one compatibilist, Daniel Dennett, explicitly likens our free actions to those of certain sophisticated machines.) However, there are today many philosophers who argue for other positions (Fischer et al. 2007). "Libertarians," in particular, argue that in some sense free actions escape the bounds of determinism and (a number at least believe) that humans ("agents") are in some sense autonomously the causes of their actions. Standing in the tradition of Bishop Berkeley (1710) – if not even that of Immanuel Kant – they strive to show how this can be, arguing that humans have "powers" to bring about events (O'Connor 2000; Clarke 2003). Here, I am not even going to try to argue for my own inclinations, because truly it does not matter to the main thread of the argument. As with the other disagreements, the dispute over the free will problem – some thinking the machine metaphor speaks to it and others denying that this is so – shows that there is no a priori determination of the scope and limits of the machine metaphor.

Four problems, three problems, seven problems. Messy-fringed problems also. Is morality simply not deducible from science (but nevertheless true in the sense of having some foundation), or is it possible to show from science that morality has no foundations at all? I make a virtue of this overall looseness to my conclusion. Science is a dynamic, powerful phenomenon. Its reach is in many respects an empirical matter. That we should at any point not know the full strength and limit of the reach is to be expected. A proof that we now have all in our grasp, about the qualities of science, would be more dubious than our present uncertainty. The important thing to take with us now is that the machine metaphor may be powerful. Yet it is not the Christian God. It is not all-powerful. It is not omnipotent.

SIX

——

ORGANICISM

Let's try one more line of thought before we turn to religion. We have traced modern science down from the Scientific Revolution of the sixteenth and seventeenth centuries. In the last chapter, this led us to conclude that there are simply a number of problems that this kind of science – science based on the machine metaphor, in some sense or another – cannot tackle – cannot even start to tackle. Is the problem with us, namely, that we have hitched our star to science of the wrong kind? The question now is not so much whether we have distorted the actual history of science – no one is going to deny that machine-metaphor science has been dominant – but whether the science itself has been on the wrong track. Should science have adopted other root metaphors and gone that way instead? Would we now find that our questions could be answered?

Really there are two questions here. First, why has the machine-metaphor kind of science predominated? Second, what alternatives are there, and do they answer our questions? Do they perhaps leave other questions – questions answered by the machine metaphor – unanswered? There are some fairly ready answers to the first question. Modern science – science since the Scientific Revolution – exhibits various epistemic virtues. It is coherent within itself and consistent across different fields or areas. In other words, it doesn't like contradictions. It unifies different areas of inquiry beneath one or a few comprehensive hypotheses – in the case of Newton's mechanics, there is no longer any need of one science for the heavens and another for the earth. Just one will do. This unification is connected to a drive for simplicity and elegance. Good science is not a ramshackle edifice of different parts but somehow has a sense of the simple nature, the purity of the world. And above all, science is predictive and predictively fertile. It lets you make forecasts about the events of the future (or, in the case of the historical sciences, unknown events of the past), and it keeps generating such predictions. It convinces you that it is not just a figment of your imagination; that it is about a real independent world of experience, that runs in a regular fashion and that hence lets you know what will be as well as what was and what is. Much of this, of course, is bound up with the technological implications of science, the ways in which it lets us control and exploit our surroundings for our own benefit.

What about alternatives – alternatives, that is, to mechanism – and why, if at all, should we take these seriously? That is the big question we must explore now and see where it leads us given our underlying interests. Of course, part of the problem here is finding alternatives. It is all very well to say that you do not like the mechanism approach, but what are you going to propose in its stead? It is hard to be definitive on these matters, but an extensive search suggests that virtually all of the alternative voices since the Scientific Revolution – since the triumph of the machine metaphor, that is – have been calls to return to the organic metaphor, in one form or another. In other words, the demand is that we again think of the world as an organism in some sense or fashion. Organicism, so called. So let us take this as our thread of Ariadne and see where it leads. I will leave comment, critical or otherwise, until our survey is finished.

NATURPHILOSOPHEN

The key starting figure is our old acquaintance, the Dutch Jewish philosopher Baruch Spinoza. In some respects, he is an unlikely figure because he would have (rightly) thought of himself as much more of a mechanist, and he certainly had no love of teleology, the mark of the organic. In a letter, he declared that he repudiated "occult qualities, intentional species, substantial forms, and a thousand other trifles" (Spinoza 1995, letter 60, to Boxel). Yet there are some puzzling passages in his writings that suggest that final causes lie close to the surface, for instance, the claim that "[e]verything, in so far as it is in itself, endeavours to persist in its own being" (Spinoza 1985, III, P, 6). More pertinent, perhaps, is the claim that God and Nature (*Deus sive Natura*) are one. Whether or not this truly implies pantheism, or whether it suggests that God, considered as a genuine being, truly does not exist, are matters we can ignore. The point is that there is more than a hint that nature, material being – Descartes' *res extensa* – and the living, even the thinking – Descartes' *res cogitans* – are one, and that means that in some sense (perhaps even in an Aristotelian sense) the whole of existence is living, is alive.

Certainly it was this that was picked up at the end of the eighteenth century, particularly in Germany by the group of philosophers, poets, writers, and theologians known as the Young Romantics (Richards 2003). The key figure for us here is the philosopher Friedrich Wilhelm Joseph von Schelling, the founder of Nature Philosophy, or (as it is generally known in the original) *Naturphilosophie*. He was explicit in saying that for him the world is organic. As we have seen, Kant, another key influence, had argued that the living world must be considered teleologically; but for the great philosopher, this was always heuristic. It was not part of the world. Schelling took the step Kant would not make, and argued that the teleology is indeed genuine – things do point to ends – and extended it to all things, living and nonliving. "Even in mere organized matter there is *life*, but a life of a more restricted kind. This idea is so old, and has hitherto persisted so constantly in the most varied forms, right up to the present day – (already in the most ancient times it was believed that the whole world was pervaded by an animating principle, called the world-soul, and the later period of Leibniz gave every plant its soul) – that one may very well surmise from the beginning that there must be some reason latent in the human mind itself for this natural belief." There is indeed a reason for this belief. "The sheer wonder which surrounds the problem of the origin of organic bodies, therefore, is due to the fact that in these things necessity and contingency are most intimately united. *Necessity*, because their very *existence* is *purposive*, not only their form (as in the work of art), *contingency*, because this purposesiveness is nevertheless actual only for an intuiting and reflective being." A term is now introduced that had a long shelf life and is indeed very popular in some circles today. Apparently, because of the notion of purpose, "the human mind was very early led to the idea of a *self* organizing matter, and because organization is conceivable only in relation to a mind, to an original union of mind and matter in these things. It saw itself compelled to seek the reason for these things, on the one hand in Nature itself, and on the other, in a principle exalted above Nature; and hence it very soon fell into thinking of mind and Nature as one" (Schelling 1988, 35).

Self-organization! The world is something that produces itself, has its developing powers inside, as an unfurling organism is driven by forces within rather than without.

> If, finally, we gather up Nature into a single Whole, *mechanism*, that is, a regressive series of causes and effects, and *purposiveness*, that is, independence of mechanism, simultaneity of causes and effects, stand confronting each other. If we unite these two extremes, the idea arises in us of a purposiveness of the whole; Nature becomes a circle which returns into itself, a self-enclosed system. The series of causes and effects ceases entirely, and there arises a reciprocal connection of *means* and *end*; neither could the individual become real without the whole, nor the whole without the individual. (pp. 40–1)

Naturphilosophie has a dreadful reputation, and much of it has been honestly earned through never-ending writings, obscure prose, and unrestricted fancy. So it is important to recognize that bringing back the organic metaphor was not simply crazy or even unconnected to the science of the day. This was the time when people were starting to take an intense interest in such phenomena as electricity and magnetism, and these are the sorts of things – especially as experiments started to show how electricity was deeply important in the workings of the organic (for instance, in things like muscle movement) – that make one think that dead matter might not be so dead after all. Particularly when you think of magnetic and electrical forces as "wanting" to find their opposites – as in north and south poles attracting each other. (Sexual metaphors here are not hard to imagine.) It is for this reason that the whole notion of polarity, and dialectic through the melding of opposites – think ultimately of Hegel and Marx – was important to the *Naturphilosophen*.

Important also was the thought of development – after all, this is what organisms do more than anything – and it was a ready inference that the world should be seen as moving up to ever-better forms. It was not long, therefore, before people started drawing analogies between the history of the individual and the history of the world, the history of life particularly. Some took this in (what we would call) a full-blooded, evolutionary fashion – Schelling most probably, perhaps also (later in life) the poet Johann Wolfgang von Goethe. For others, it was more the idea that counted (Ruse 1996).

> Nature is to be regarded as a *system of stages*, one arising necessarily from the other and being the proximate truth of the stage from which it results; but it is not generated *naturally* out of the other but only in the inner Idea which constitutes the ground of Nature. *Metamorphosis* pertains only to the Notion as such, since only its alteration is development. But in Nature, the Notion is partly only something inward, partly existing only as an individual; *existent* metamorphosis, therefore, is limited to this individual alone. (Hegel 1817, 21)

There was one more very important element to the mix. Although it does seem fair to say that the teleology of the *Naturphilosophen* was more Aristotelian than Platonic – that is, that it was more of an internal teleology, seeing the forces as part of the system, than an external teleology, seeing design imposed on the system by an outside intelligence – in thinking of the world as organic, one is clearly harking back to the Plato of the *Timaeus*. But once we hint at a Platonic lineage, then we start to think also in terms of underlying Forms or Ideas, and this obviously ties in with what Kant had noted (Aristotle too, for that matter), namely, that organisms seem to have been built on certain underlying patterns or archetypes. The similarities between the skeletons of vertebrates speak to this insight (Russell 1916). It was but a moment's work for the *Naturphilosophen* to seize on this notion, seeing repeated basic ideas not only across organisms, but also within organisms, and also linking organisms with the inorganic – for instance, with the patterns of snowflakes (something noted carefully by Kant in the *Third Critique*) – as the basic ideas get repeated.

Particularly influential here was the theory of the vertebrate skull, supposedly first formulated by Goethe after looking at a dried sheep's skull in the Jewish cemetery in Venice and also independently (and according to him, really first discovered) by the anatomist Lorenz Oken. If you look at the skull, you can see sutures, and Goethe and

Oken argued that these suggest that the skull is in fact part of the backbone, the various pieces being modified vertebrae. In a sense, therefore, it could be argued that nature took one piece, modified and repeated it in order to make the ideal vertebrate body, and then used this body as a pattern for all other real bodies, which revealed themselves through time in ever-increasing perfection – ending, of course, with *Homo sapiens*.

Somewhat anachronistically using today's language, this picture was antireductionist and antimechanist. It was organicist and wholist – or, as it is usually put, holist. By this is meant that the key to understanding nature is to see it as a related system of parts, none of which makes any sense in isolation, and which together yield things that do not exist apart – the very fact of life, for a start. It is the integration that counts, the working to ends, rather than the turning of cogs and wheels.

LATER DEVELOPMENTS

We see the difference from the picture of the past chapters when we move to Britain later in the century. On the one hand, we have the machine metaphor–influenced Charles Darwin, who is proposing physical mechanisms (natural selection, most prominently), who is trying not to build progress into the picture as an inevitable outcome, and who explicitly uses mechanistic heuristics to understand the functioning of adaptation. He was aware of the similarities between organisms, but he thought them a straightforward consequence of evolution through selection, and he wanted no part of transcendental ideas. Although at one point he adapted the vertebrate theory of the skull, when Huxley (1857–59) tore it apart on anatomical grounds, he dropped it like a hot potato.

On the other hand, we have Richard Owen, a brilliant anatomist and paleontologist, inclined to evolution (for all that after the *Origin* he was associated with the opposition to Darwin) but even more inclined to *Naturphilosophie*. He saw the whole of life as an upward progression, and thought that it exhibits archetypes, things he likened specifically to Platonic ideas. For him, it was all a matter of the ways in which Nature unfurls herself organically.

> The Archetypal idea was manifested in the flesh, under divers such modifications, upon this planet, long prior to the existence of those animal species that actually exemplify it.

To what natural laws or secondary causes the orderly succession and progression of such organic phenomena may have been committed we as yet are ignorant. But if, without derogation of the Divine power, we may conceive the existence of such ministers, and personify them by the term 'Nature,' we learn from the past history of our globe that she has advanced with slow and stately steps, guided by the archetypal light, amidst the wreck of worlds, from the first embodiment of the Vertebrate idea under its old Ichthyic vestment, until it became arrayed in the glorious garb of the Human form. (Owen 1849, 86) [Figure 6.1]

In the post-*Origin* era, people like Thomas Henry Huxley in Britain and Ernst Haeckel in Germany were strong mechanists – we have seen this already with Huxley – but the organicist philosophy had left its mark. The scientific thinking of the day stressed homologies and progress and links between individual development and group change. That was the essence of the "biogenetic law," ontogeny recapitulates phylogeny (Haeckel 1866; Richards 2008). In Germany, this kind of thinking in evolutionary biology has continued to the present, but in the Anglophone world with the coming of the synthetic theory of evolution – the synthesis of Darwin and Mendel in the 1930s – it played little role in evolutionary thought. However, in the past twenty or thirty years, particularly thanks to the championing of the paleontologist and popular-science writer Stephen Jay Gould, it has had a renaissance. In his famous article "The Spandrels of San Marco," cowritten with a fellow Harvard academic, Richard Lewontin, Gould decried the reductionistic mechanism of conventional Darwinian evolutionary theory and urged a return to a more holistic, archetype-based view of organic nature. Gould and Lewontin wrote with warmth of the German tradition, which "denies that the adaptationist programme (atomization plus optimizing selection on parts) can do much to explain *Baupläne* [archetypes] and the transitions between them. But it does not therefore resort to a fundamentally unknown process. It holds instead that the basic body plans of organisms are so integrated and so replete with constraints upon adaptation... that conventional styles of selective arguments can explain little about them" (Gould and Lewontin 1979, 160). Conventional Darwinism has "underrated the importance of integrative developmental blocks and pervasive constraints of history and architecture" (p. 163).

Gould's very public stand encouraged others to come out of the dark. Pushing his line of argumentation to the extreme, we now

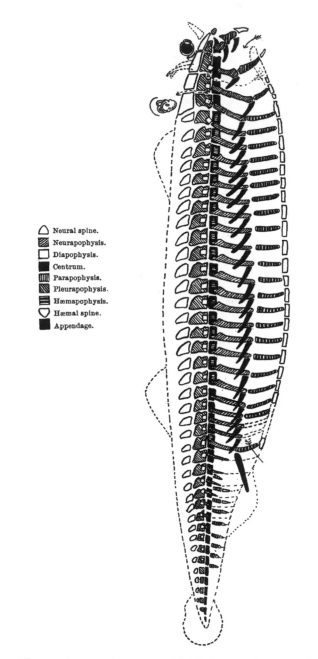

Neural spine.
Neurapophysis.
Diapophysis.
Centrum.
Parapophysis.
Pleurapophysis.
Hæmapophysis.
Hæmal spine.
Appendage.

Figure 6.1. The vertebrate archetype on which living vertebrates are based, from Richard Owen, *On the Nature of Limbs* (1849).

find enthusiasts for self-organization arguing that we get "order for free," a position strongly redolent of *Naturphilosophie*. Enthusiasts argue that much of the complexity that we find in nature – including organic nature – flows naturally from its very essence, without the need of supplements like natural selection. Nature itself, including inorganic nature itself, has its own principles of ordering and complexifying and growth. The computer-intoxicated theoretical biologist Stuart Kauffman writes: "The tapestry of life is richer than we have imagined. It is a tapestry with threads of accidental gold, mined quixotically by the random whimsy of quantum events acting on bits of nucleotides and crafted by selection sifting. But the tapestry has an overall design, an architecture, a woven cadence and rhythm that reflects underlying law – principles of self organization" (Kauffman 1995, 185).

In like fashion, the biologist Brian Goodwin (2001) draws our attention to something known as the Beloussov-Zhabotinsky reaction, so named after a couple of Moscow-based scientists of the 1950s. When organic and inorganic liquids are placed on a flat plane (as in a Petri dish), they go through a kind of ring-making exercise that moves out concentrically to the edges. These are very much the kind of movements that one sees in nature itself, particularly in the slime molds, which behave this way when food gets scarce. Goodwin writes: "Cells start to signal to one another by means of a chemical that they release. This initiates a process of aggregation: the amoebas begin to move toward a center, defined by a cell that periodically gives off a burst of the chemical that diffuses away from the source and stimulates neighboring cells in two ways: (1) cells receiving the signal themselves release a burst of the same chemical; and (2) the move toward the origin of the signal" (Goodwin 2001, 46). As the amoebas go through their movements, leading ultimately to union and then to fruiting and the production of more amoebas, their paths are exactly those found in the Beloussov-Zhabotinsky reaction.

The molecules involved in the two cases are quite different, but obviously the feedback systems are parallel. And it is all a question of the way in which nature itself has powers of organization. Defining a field as "the behavior of a dynamic system that is extended in space," Goodwin concludes:

> A new dimension to fields is emerging from the study of chemical systems such as the Beloussov-Zhabotinsky reaction and the

Figure 6.2. A Beloussov-Zhabotinsky pattern.

similarity of its spatial patterns to those of living systems. This is the emphasis on self-organization, the capacity of these fields to generate patterns spontaneously without any specific instructions telling them what to do, as in a genetic program. These systems produce something out of nothing.... There is no plan, no blueprint, no instructions about the pattern that emerges. What exists in the field is a set of relationships among the components of the system such that the dynamically stable state into which it goes naturally – what mathematicians call the generic (typical) state of the field – has spatial and temporal pattern. (pp. 51–2) [Figure 6.2]

Striking right home at the Darwinians, people like Goodwin seize on supposedly paradigmatic examples of natural selection in action, arguing that in fact they are no more (or less) than the unfurling patterns of nature itself. Take phyllotaxis, the patterns shown by many flowers and fruits in the plant world, for instance, the spirals of the sunflower or the twisting lines shown by pine cones [Figure 6.3]. Darwinians have long argued that these are of direct adaptive significance, usually associated with maximizing the amount of sunlight that falls on some specific seed or other. This was the position of Chauncey Wright, an early American pragmatist and great supporter of Darwin. "To realize simply and purely the property

Figure 6.3. A sunflower showing phyllotaxis.

of the most thorough distribution, the most complete exposure to light and air around the stem, and the most ample elbow–room, or space for expansion in the bud, is to realize a property that exists separately only in abstraction, like a line without breadth" (Wright quoted in Gray 1881, 125).

Organicists like Goodwin, however, seize upon the mathematics of the case. The ways of growth force the components into certain familiar grids or lattices, and these in turn are amenable to fairly simply mathematical analysis. The plants in question produce their parts from the center and then push out as they grow. In a sunflower, for instance, one gets one seed and then another and then another – this produces the genetic spiral. As the seeds line up, one by one, new lines or patterns emerge – the most noticeable spirals are known as parastichies. The seeds running along any particular spiral, numbering them in the order they were produced, exhibit fixed patterns. Remarkably, the numbers from the crisscrossing parastichies have a formula. The differences between the seeds going the one way (clockwise) and those going the other way (counter clockwise), follow the sequence: 0, 1, 1, 2, 3, 5, 8, 13,... This is the formula worked out by the thirteenth-century Italian mathematician Leonardo Fibonacci, who tried to calculate the number of descendents in any generation

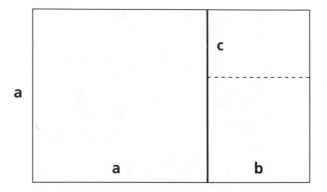

Figure 6.4. To get the Golden Mean, you divide a by a + b, which in turn equals b divided by b + c, where b + c equals a. Hence a/(a + b) = b/a, and if you set a = 1, this yields the Golden Mean for b.

from an initial pair of breeding rabbits. The formula is of course better known today as one of the clues in the thriller *The Da Vinci Code*.

Goodwin argues that that is all there is to it. The developing plants simply follow the rules of mathematics, and biological forces have nothing to do with anything. Goodwin is not just Platonic but practically Pythagorean in his numerological enthusiasms. The vulgar fraction series formed by dividing successive members of the Fibonacci series homes in on 0.618, which in turn is what the ancient Greeks called the Golden Mean, the figure arrived at by dividing the sides of a rectangle such that removing a square from the rectangle leaves one with a smaller but identical rectangle [Figure 6.4]. As it happens, you can get the Golden Mean out of circles, too, if you divide up the perimeter properly. This gives you a major angle of 137.5 degrees, which (and if you are not yet convinced, you will be now!) is just the angle on the genetic spiral that divides successive leaves or parts. "So plants with spiral phyllotaxis tend to locate successive leaves at an angle that divides the circle of the meristem in the proportion of the Golden Mean. Plants seem to know a lot about harmonious properties and architectural principles" (Goodwin 2001, 127). (The meristem is the growing tip of the plant.) [Figure 6.5].

EMERGENCE

The *Naturphilosophen* would feel very much at home with this kind of thinking. Not that they have been the only people ever to think

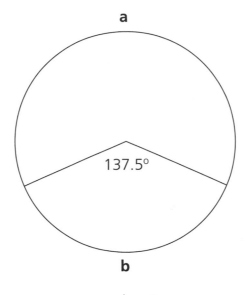

a + b = 1

Figure 6.5. The Golden Mean is a divided by a + b.

in these sorts of ways. The extent to which the *Naturphilosophen* can be considered out-and-out vitalists, meaning people who subscribe to some kind of Aristotelian force governed by final causes, is much debated (Lenoir 1989; Richards 2003). Kant and Schelling were both influenced by the claims of the biologist Johann Friedrich Blumenbach (1781) about that thing he called a *Bildungstrieb*, which is a force that governs and makes possible organization. Yet Kant (to pick up on discussion earlier in this book) was probably not a hard-line vitalist, because he deliberately denied that he was talking about the real world. He was talking about the way we conceive it. Schelling certainly pushes us a lot closer to the edge, although as always one must think of the force as a kind of organizing principle rather than as a thing like a chair or a table – or as a proximate cause, like the force one uses to open a jammed door.

As we know, vitalism was a hardy plant. Later in the nineteenth century there were vitalists, and by the end of the century the idea was positively thriving, what with the embryologist Hans Driesch with his entelechies and the philosopher Henri Bergson with his *élans vitaux*. But, as has also been pointed out, the trouble with vitalism is that it really does not do very much – at least, it does not help much in the satisfaction of the prized epistemic virtues of science. Belief in vital forces does not make it any easier to predict things or to unify different fields. The twentieth century saw several movements that tried to keep with the sympathies of vitalism, and to repudiate hard-line mechanism, without necessarily falling into the trap of being useless or in some sense metaphysically unacceptable. One movement was that of (the already-mentioned) holism, a term coined by the South African statesman Jan Smuts. He defined it as: "The tendency in nature to form wholes that are greater than the sum of the parts through creative evolution" (Smuts 1926). This is very much an Aristotelian notion – "the whole is more than the sum of its parts" (*Metaphysics* 1045a10) – and it can be seen how it readily lends itself to some kind of organicism, since the functioning living organism is often felt to be more than just the bare molecules of which it consists.

The twentieth-century physicist David Bohm was sympathetic to this philosophy, arguing that in modern physics the observer and the observed are no longer two different things – trying to measure at the quantum level changes radically the thing being measured – and that hence we ought to have a philosophy that is more directed to the whole than simply to the disconnected parts. Speaking of the paradoxes of quantum mechanics, he suggests that our problems are a function of the ways in which we try to break things down in a reductive sense, looking for the functionings of the parts.

> Questions of this nature cannot be met properly while we are caught up, consciously or unconsciously, in a mode of thought which attempts to analyse itself in terms of a presumed separation between the process of thinking and the content of thought that is its product. By accepting such a presumption we are led, in the next step, to seek some fantasy of action through efficient causes that would end the fragmentation of the content while leaving the fragmentation in the actual process untouched. What is needed, however, is somehow to grasp the overall *formative cause* of fragmentation, in which content and actual process are seen together, in their wholeness. (Bohm 1980, 18)

Closely related to holism is "emergentism," a term coined by George Henry Lewes (the philosopher and longtime lover of George Eliot) in 1875. This too is a whole-is-greater-than-the-parts position, arguing that there are levels of experience and that somehow the lower levels can never fully account for the upper levels – as one might, say, expect the discussion of the gears and pullies in a mill to explain why it is that the wheels go round, and what they do when they do. John Stuart Mill was a paradigmatic emergentist. Here he is on the subject of life:

> All organised bodies are composed of parts, similar to those composing inorganic nature, and which have even themselves existed in an inorganic state; but the phenomena of life, which result from the juxtaposition of those parts in a certain manner, bear no analogy to any of the effects which would be produced by the action of the component substances considered as mere physical agents. To whatever degree we might imagine our knowledge of the properties of the several ingredients of a living body to be extended and perfected, it is certain that no mere summing up of the separate actions of those elements will ever amount to the action of the living body itself. (Mill 1843, Book III, Chapter 6, §1)

Generally emergentists deny energetically that they are vitalists. They are not supposing new forces but rather that principles of organization, operative only at upper levels, produce new things and sensations and so forth. In the early twentieth century, the philosophers C. D. Broad (1925) and Samuel Alexander (1920) were big enthusiasts for the idea. Broad saw his position as explicitly opposed to mechanism, arguing that there is need of "trans-ordinal laws" that take you from one level of organization to another. "A trans-ordinal law would be a statement of the irreducible fact that an aggregate composed of aggregates of the next lower order in such and such proportions and arrangements has such and such characteristic and non-deducible properties" (1925, pp. 77–8).

Usually what such thinkers have directly in mind is life itself and then consciousness. Samuel Alexander was explicit:

> Physical and chemical processes of a certain complexity have the quality of life. The new quality life emerges with this constellation of such processes, and therefore life is at once a physico-chemical complex and is not merely physical and chemical, for these terms do not sufficiently characterize the new complex which in the course and order of time has been generated out of them. Such is the account to be given of the meaning of quality as such. The higher quality emerges from the lower level of existence and has its roots therein, but it emerges therefrom, and it does not belong to that level, but constitutes its possessor a new order of existent with its special laws of behaviour. The existence of emergent qualities thus described is something to be noted, as some would say, under the

compulsion of brute empirical fact, or, as I should prefer to say in less harsh terms, to be accepted with the "natural piety" of the investigator. It admits no explanation. (Alexander 1920, 2, 46–7)

He felt the same way about consciousness, calling it "something new, a fresh creation." What it is not is something "merely neural."

Today emergentism is having something of a philosophical renaissance, although not without controversy (Clayton and Davies 2006). Most people do not much want to deny some weak form of emergence, a form sufficiently weak as not to clash with mechanism. If, for instance, one says that Katrina devastated New Orleans, one is talking about a hurricane. But most would agree that this does not mean a great deal more than wind and rain and so forth. Katrina did things that one raindrop and one puff of air could not do, but there is nothing vitally new about Katrina as a whole. Natural selection is another thing that most would say is a meaningful notion – it is a force or mechanism – but at another level it is just differential reproduction. Some organisms because of their features do better at reproduction than others. That is all there is to it. Nothing more.

The tough questions come when you start to talk about stronger versions. David Chalmers (2006), whom we met in the last chapter, argues that there is probably only one case of strong emergence, namely, consciousness. Others – the physicist Paul Davies (2006, 199), for instance, would certainly include life. And yet others would go much further. One suspects that many of the self-organization enthusiasts would fall into this camp. One who certainly does is the linguist Terrence Deacon. He writes:

> Unlike the logic of machine design, however, in which things must be forced to occur, pushed into place, and restricted in their deviant tendencies, the logic of organism 'design' instead depends on recruiting the spontaneous intrinsic tendencies of molecular substrates and structural geometries. A superficial interpretation of the evolutionary process might suggest that the introduction of new ordering principles in organism design derives solely from lucky accidents achieving serendipitous functions. Yet an analysis of intracellular molecular processes and of the morphogenetic mechanisms of development suggest that a significant fraction of the order–generating processes of life are instead due to self-organizing dynamics, which are intrinsic to molecular geometries and cell-cell communications. Indeed, the majority of macromolecular structures in cells tend to self-assemble, and the majority of

critical chemical reactions occur within self-reinforcing cycles of reactions. Machines need to be built with extrinsic means, but organisms must develop themselves. Where there is no external means for the generation of order, order must arise from tendencies already present. (Deacon 2006, 116)

As suggested by people like Stuart Kauffman and Brian Goodwin, apparently the basic processes of the universe can generate order, up to and including organisms. There seems to be a seamless whole from the inorganic to the fully living. This may not be vitalism, but it is surely in the same family.

GAIA AND ECOFEMINISM

Let us now turn briefly to those who are pushing the organic metaphor less for philosophical reasons and more because they think it has greater social traction. Most obviously these are the people who decry the rape of our planet – lost rain forests, pollution in the big cities, global warming, fields turned to deserts by the demands of overpopulation, seas empty of fish and other denizens – and who think that our mechanistic philosophy somehow is to blame. At the center of this charge is the Gaia hypothesis, the brain child of the British scientist James Lovelock (1979), aided by the American microbiologist Lynn Margulis (Margulis and Sagan 1997). Lovelock argues outright that the earth is a living organism, that it maintains itself in a kind of homeostasis, with a feedback that keeps the key constants (like the salinity of the oceans) in balance. It was apparently the trips to space, where the astronauts could look back at earth, that triggered this realization. "There is nothing unusual in the idea of life on Earth interacting with the air, sea and rocks, but it took a view from outside to glimpse the possibility that this combination might consist of a single giant living system and one with the capacity to keep the Earth always at a state most favorable for the life upon it" (Lovelock 1979, 308).

The trouble is that the feedback can do only so much, and humans have knocked things out of balance. Fortunately, we are now starting to realize this. "Most of us sense that the Earth is more than a sphere of rock with a thin layer of air, ocean, and life covering the surface. We feel that we belong here as if this planet were indeed our home. Long ago the Greeks, thinking this way, gave to the Earth the name

Gaia or, for short, Ge. In those days, science and theology were one and science, although less precise, had soul" (Lovelock 2007, 307). Interestingly, it seems not to have been the mechanists as such who spoiled all of this. It was the medieval thinkers – a point one might have wished to see spelled out in more detail, given the debt that these thinkers had to Aristotle. But no matter, the damage was done. "As time passed this warm relationship [between science and theology] faded and was replaced by the frigidity of the schoolmen. The life sciences, no longer concerned with life, fell to classifying dead things and even to vivisection. Ge was stolen from theology to become no more than the root from which the disciplines of geography and geology were named" (ibid.). Fortunately, what goes around comes around. "Now at last there are signs of a change. Science becomes holistic again and rediscovers soul, and theology, moved by ecumenical forces, begins to realize that Gaia is not to be subdivided for academic convenience and that Ge is much more than just a prefix" (ibid.).

So, where does this all lead and leave us?

> If we are "all creatures great and small," from bacteria to whales, part of Gaia then we are all potentially important to her well being. We knew in our hearts that the destruction of a whole range of other species was wrong but now we know why. No longer can we merely regret the passing of one of the great whales or the blue butterfly, nor even the smallpox virus. When we eliminate one of these from Earth, we may have destroyed a part of ourselves, for we also are a part of Gaia. (p. 308)

Lovelock concludes that we humans are the senses and nervous system of Gaia. Hence: "Through our eyes she has for the first time seen her very fair face and in our minds become aware of herself. We do indeed belong here. The earth is more than just a home, it's a living system, and we are part of it" (ibid.).

Following on his early writings, Lovelock was accused of (among other things) an explicitly teleological stance, involving mushy final causes. Countering, he has devised his so-called daisy-world model (Watson and Lovelock 1983). Suppose you have two kinds of daisies covering a planet. One kind is black and absorbs heat but does not do very well when things are really hot. The other kind is white and reflects heat and does not do very well when things are really cold. The sun shines on the rather cold planet. The black plants initially do rather well, but they absorb heat, and before long they not only cover most of the planet, they heat it up. Now the white plants are at a selective advantage, and so they take over. The point is that however you interfere with the planet – short of killing off all the plants of one kind – it has a built-in mechanism to right itself and find a natural balance. There is no unwonted teleology here, although there is certainly the appearance of such.

Sympathetic to this kind of organic thinking are many who subscribe to the various ecological movements – using the term "ecological" now in the sense of those with burning moral concerns and less in the sense of a formal science (although not thereby intending, in any condescending way, to separate the two). One finds supporters of GreenPeace, for instance, and those who decry genetically modified foods, and (to a certain extent overlapping) particularly those who label themselves "ecofeminists." Prominent here has been the historian of science Caroline Merchant. Her detailed and (to be fair) rather thrilling book, *The Death of Nature* (1980), tells how the organic metaphor ruled until the Scientific Revolution, when it was overtaken by the machine metaphor. Nature was no longer considered living, although, thanks to the powerful metaphors of the philosopher Francis Bacon, it was considered feminine in the sense that it could be violated and raped by the inquiring male scientist.

> Sensitive to the same social transformations hat had already begun to reduce women to psychic and reproductive resources, Bacon developed the power of language as a political instrument in reducing female nature to a resource for economic production. Female imagery became a tool in adapting scientific knowledge and method to a new form of human power over nature. The "controversy over women" and the inquisition of witches – both present in Bacon's social milieu – permeated his description of nature and his metaphorical style and were instrumental in his transformation of the earth as a nurturing mother and womb of life into a source of secrets to be extracted for economic advance. (Merchant 1995, 80)

For Merchant and those who think like her, prominently the Indian activist Vandana Shiva (2000, 2005), we live still with the effects of this kind of thinking. The world is there to be exploited, with no understanding of its worth in its own right – its worth as a living, female being. But now, thanks to the writings of people like Rachel Carson, whose *Silent Spring* detailed the damage done to the ecosystem by such artificially introduced substances as DDT, ecofeminists of both sexes are starting to realize that we must reassess and take a different path, one much more in tune with the earth as a living being. "If an earthly paradise or even a small homeplace on the larger garden planet were possible, it would retain wilderness within a rural setting and maintain natural and cultural diversity in all its forms." Like homesteaders in a cowboy movie, we will be living in little log cabins, digging and planting in the fertile soil, surrounded by looming but friendly mountains, soothed by gentle breezes, listening to the bubbling of trout-filled streams. Although there is some question about whether fishing will be permitted. "Once in a while a trout rises but is not taken; my creel is empty, but my mind is full. Heaven is on Earth" (Merchant 2003, 203).

UNASKED QUESTIONS?

We now have before us a representative sampling of the kinds of attacks that have been made on the machine metaphor since its triumph in the sixteenth and seventeenth centuries, and of the kinds of arguments that are made for the restoration of the organic metaphor. Our next task must be to see if the organic metaphor, like the machine metaphor, leaves questions unasked and unanswered, and if so, what is their nature? Are they identical to the unasked questions for the machine metaphor?

Our general discussions about metaphor tell us that there will undoubtedly be limits. Let us therefore start by running through those limits that were listed and discussed in the last chapter. First, why is there something rather than nothing? At one level, under the organic metaphor, I suppose one might say that this really is a meaningless, or at least a pointless, question. The world is an organism, and it persists and exists because that is what functioning, living organisms do. I am not sure that anyone is thinking in terms of the world reproducing itself, and if they are, it really does not enter into the main question. Nonetheless, there is often the thought – there was certainly the thought for the *Naturphilosophen* – that the world is growing and developing in a direction of maturity. But whatever refinements you want to put on the picture, ultimate questions don't really enter in. This, remember, was a point made by Adolf Grünbaum.

However, at another level, both the scientific and the metaphysical critic will claim that this is really no true answer to the question. One presumes for a start that the organic metaphor enthusiasts believe that the earth is only 4.5 billion years old and that the Big Bang occurred 15 billion years ago, in which case exactly the same question arises as for the mechanical metaphor. I suppose one could get into some sort of super-organism view of the universe, with the earth truly being an offspring in a sense. (Are the other planets also living, or are they stillborn?) But even with this picture, one still has the sense of the question of what makes anything exist at all – even if the physical world is infinitely old. This sense of the question is no more touched by the organic metaphor than it is by the mechanical metaphor. It is possible that, picking up on the point made by Adolf Grünbaum, you might object that back with the Greeks, when the organic model flourished, no one asked questions about ultimate origins. But even if this is true, we are not back with the Greeks, and there seems to be no a priori reason why we should not ask it. And on this, the organic model is silent.

The question of values is an interesting one. Organicists claim they can derive values from the way things are. "Life, and with it agency, came naturally to exist in the universe. With agency came values, meaning, and doing, all of which are as real in the universe as particles in motion" (Kauffman 2008, x). The whole point about the world as an organism is that it is not dead and lifeless, without any value of its own. "Agency, values and 'doing' did not come into being separately from the rest of existence; they are emergent in the evolution of the biosphere" (p. 8). The world is living and as such has a value in itself. This point was stressed earlier when looking at the Greek use of the organic metaphor. Likewise here. If, for instance, Gaia be true, then there are good reasons why you should not treat the world in a way that harms or destroys it. It is your mom, and you don't treat moms – particularly not your own mom – this way. As Caroline Merchant puts it about the pre-mechanistic world picture, the one she hopes to recreate: "The image of the earth as a living organism and nurturing mother served as a cultural constraint restricting the actions of human beings. One does not readily slay a mother, dig into her entrails for gold, or mutilate her body. As long as the earth was conceptualized as alive and sensitive, it could be considered a breach of human ethical behavior to carry out destructive acts against it" (Merchant 2003, 43).

Many working with this metaphor would argue that the world increases in value. This would seem to be the position of the *Natur-philosophen*. Just as the organism matures, so also the world has matured, and the end result is better than the beginning. In particular, we now have humans, and they are at the top of the chain of being. In a way, this would also seem to be the attitude of Lovelock – we have seen how he thinks of humans as the senses and the thinking apparatus of the world. You might argue that people like the ecofeminists assume that humans – males, specifically – are evil, given the rape of the environment that has been the pattern of the past five centuries. But even they see redemption through the human. It is not the rain forests of the Amazon that are going to prevent global warming, but sensitive and aware humans. No doubt some would prefer a world entirely peopled by the female, but as Darwinians point out, often you have to be satisfied with the actual rather than the ideal.

If you are working within the machine metaphor, then this kind of thinking is fallacious. You are riding through Hume's law, deriving "oughts" from "is." But the point is that now you are not working within the machine metaphor. You are working within a metaphor that starts with value, the value of the organism that is earth. At one level the value is basic; there is no authority or foundation behind it. At another level, as we have just seen, even the organicist might ask the question about the ground of being, and this surely inserts a place for value. Why is there something rather than nothing, and why is this something a thing of great value? Why was the living being a fertile and loving being, a mother, rather than a rotten tomato? Why is it almost surely a mammal, which suckles its young, rather than a virus or a vile parasite? So my suspicion is that ultimately the organic metaphor, no less than the machine metaphor, leaves room for value questions. There are things to which it simply does not speak.

Mind, in a way, is in the same position as value. If you think about it, people like Leibniz and David Chalmers do have a point. If you argue that the world is a dead substance, an eternally whirling machine, then it is hard to see any place for mind. Dualism may not be the solution, but to an extent you have already forced it upon yourself. There is no mind in the world, *res extensa*, so you have to add it as a different kind of extra, *res cogitans*. Really, the only alternative is to go the route of people like the Churchlands and to argue that you can get mind out of dead matter. You may not much care for their solution, but given the dilemma – given the metaphysics has been forced upon you by the machine metaphor – you really do not have much choice. At first glance, it may seem silly to go this way, but the more you think about it, the more you have sympathy for those who do go this way – and respect for their courage in going this way.

Nevertheless, one can forgive organicists for being rather smug. For them, everything is living – as it was for Aristotle. Of course, as also for Aristotle, not everything is fully conscious, but some aspects of the world – for we humans are part of the world – are conscious. In a way, consciousness is part of the organic-metaphor package. It comes with the territory, in a fashion that it does not come with the territory of the machine metaphor. And as we have just seen, organicists have a good idea of how consciousness comes about. It emerges as a higher level of being. There are levels of complexity, levels of existence, all of the way up, yielding new features, new properties, new entities, that do not exist at lower levels. "The higher quality emerges from the lower level of existence and has its roots therein, but it emerges therefrom, and it does not belong to that lower level, but constitutes its possessor a new order of existent with its special laws of behaviour" (Alexander 1920, 2, 46). Consciousness is the cream on the top. David Chalmers writes: "I think there is exactly one case of a strongly emergent phenomenon, and that is the phenomenon of consciousness. We can say that a system is conscious when there is something it is like to be that system; that is, when there is something it feels like from the system's own perspective. It is a key fact about nature that it contains conscious systems; I am one such. And there is reason to believe that the facts about consciousness are not deducible from any number of physical facts" (Chalmers 2006, 246).

I want to stress that I do not see this as a stupid position to take. Indeed, it is admirably brave in wanting to tackle the consciousness issue head on. However, in an important way it really does not do very much. You still have the question of why consciousness emerges. Indeed, after a fashion you are making it more mysterious. It is something that could not have been predicted and is impossible to explain, at least to explain fully – otherwise, why the talk of emergence? In short, even though (with reason) organicists may feel that they locate consciousness more comfortably within their systems, at another level it is as unexplained as it is for the mechanist – with the reductionist approach of people like the Churchlands ruled out as impossible. (Even mechanists may not think this ruling a great handicap to understanding.)

To coin a phrase, a familiar pattern is starting to emerge. Turn to the matters of unanswerable questions and ultimate purposes. In one sense, the organic metaphor has no unanswerable questions and, being frankly and openly teleological in a way that the machine metaphor is not, can speak to ultimate purposes. There are no questions outside the maintenance and well-being of the organism that is earth, and this maintenance and well-being are the ultimate purpose of everything. The earth is good, and it is our obligation to cherish and protect this earth, our mother. There is nothing beyond this. There could be nothing beyond this, because it is from the earth that all value derives. At another level, of course, one might ask about the ultimate meaning of it all and whether this is a question we could ever answer. It links back to why the earth exists at all. Is there any real purpose to the earth? Can it be that its own self-being is all there is to things, or is the earth in some sense fulfilling a bigger destiny? The organic metaphor can answer this sort of question as little as can the machine-metaphor. Although one suspects that many organicists will be unperturbed at our pointing out these gaps in explanation – let us not call them failures – with our understanding of the function of metaphor we would be surprised if they did not exist. That is how metaphors work.

Finally, for completeness, let us move on briefly to the other questions, questions that I feel are now explicable on the machine-metaphor position. What about the question of life itself? What about free will? As far as life is concerned, my suspicion is that most organicists would take a step back from the confident enthusiasm of the mechanists. Or rather, they would be inclined to say that the mechanists have no firm basis on which to ground their enthusiasm and confidence. Their position generally seems to be that life itself is an emergent. This was certainly the position of the older emergentists. Remember John Stuart Mill. Likewise Broad and Alexander. "To call [a structure] organism is but to mark the fact that its behaviour, its response to stimulation, is, owing to the constellation, of a character different from those which physics and chemistry are ordinarily concerned with, and in this sense something new with an appropriate quality, that of life" (Alexander 1920, 2, 62). It is also the position of many of today's emergentists. Paul Davies has written of new laws of organization being needed. Terrence Deacon writes: "Even if . . . there is unbroken causal continuity across the threshold from non–life to life, machine to mind, we nevertheless require an explanation for why causal architecture changes so abruptly at these transitions and why it is so difficult to follow the logic linking human teleological experience with its physical basis" (Deacon 2006, 113).

A passage like this shows both the strength and the weakness of the emergentist position. On the one hand, it leads you to expect things like the emergence of life. It is part of the way in which, as you go up the scale, new entities and ways of being emerge. On the other hand, it does not really tell you why life emerges, and – if emergence theory is to have any real bite – it *cannot* tell you, in some sense. Life is an emergent! End of questioning. Which may be all you can say within the scientific context but still leaves open metaphysical questions about the coming of life. Why does life emerge? This is still a meaningful question, even for the organicist, even if there is no ultimate scientific answer. If the emergentist claims that there can be a scientific answer, then the mechanist rightly can ask: What then is the difference between an emergentist and a mechanist? Remember, most mechanists are comfortable with a weak form of emergence. If it is strong emergence at issue here, then the mechanist can rightly ask: What it is that escapes mechanistic explanation, and what, if anything, can the emergentist do other than point to it?

I suspect that most organic-metaphor supporters, having rejected mechanism, are not sympathetic to a Humean compatibilist explanation of free will. They want something stronger in its place. George Ellis, for instance, writes of humans as being "autonomous agents pursuing goals proceeding from within their own thought" and of "conceiving oneself as an intelligent free agent" (Ellis 2006, 96). Linking consciousness to quantum mechanics, Stuart Kauffman suggests that this hypothesis might be "a conceivable first wedge" that might crack the free will problem. Apparently "a partially quantum conscious mind might be neither deterministic, nor probabilistic. It just conceivably might be partially beyond natural law" (Kauffman 2008, 198). It is certainly the position of some (all?) such thinkers that freedom itself is an emergent property. In a way, for these people, this is something that almost comes with the territory of thinking of mind as an emergent. If mind is something that transcends the purely physical in some sense, something that is conscious or sentient in a way that brute matter is not, then freedom is part of that very consciousness and so is itself an emergent. Indeed, many of today's emergentists talk of "top-down causation," meaning that the new upper levels of being can rebound down the scale and affect lower levels – causation is not simply a one-way process from small to big, from simple to complex (Murphy 2006). Certainly the ways in which thinking beings chose to do things with and in the world is a case of top-down causation, and freedom is very much bound up with this (Murphy and Brown 2007).

However, organicists are not without their own problems in the free will arena. This was particularly true of the *Naturphilosophen*. They rejected the noumenal world of Kant, the thing in itself, the place where Kant located our radical freedom – something that lies beyond the realm of the phenomenal, the world of Newtonian causes. As we have seen, they harked rather back to Spinoza. But in an important sense, Spinoza denies freedom! It is God or nature, everything working according to fixed, immutable laws. "The laws and rules of nature, according to which all things happen, and change from one form to another, are always and everywhere the same. So the way of understanding the nature of anything, of whatever kind, must also be the same, viz. through the universal laws of nature" (Spinoza 1985, Preface to Part III).

The only freedom one has here (as noted, something not entirely different from Hume's conception) is the freedom of nature inasmuch as it is self-sustaining, with nothing from outside forcing or interfering, and we humans are free inasmuch as we are part of this nature. This is the kind of position toward which the early Romantics edged. Only nature has some kind of absolute freedom, but inasmuch as we are part of that nature, we too are given a sort of freedom. But don't forget that we are right at the top of creation. That gives us a higher dimension of freedom than, say, a rock or a cabbage. "The first idea is naturally the representation of myself as an absolutely free being. With this free, self-conscious being an entire world arises simultaneously – out of nothing – the only true and conceivable creation out of nothing" (Richards 2003, 124–5, quoting Schelling). It is not entirely silly or cynical to suggest that for the Young Romantics, sexual passion and love were the highest expression of human freedom, and certainly not something that is just the result of blind law.

To be honest, people like today's ecofeminists are generally less interested in arguing the philosophical niceties of the notion of freedom and more in assuming a fairly robust sense and then exhorting us to action – an action that depends on our having the ability to choose the right course of conduct for ourselves. So let us leave this discussion sensing that here too the organic metaphor does not offer a full solution simply based on scientific grounds.

ORGANICISM SUMMED UP

Let me make one point clear. I have not been offering the alternatives of this chapter in order to get you to jettison the machine metaphor and turn back to the organic metaphor. You may wish to, and that is your business. My aim, rather, is to strive for relative completeness. I do not want critics charging that I am making my points only by ignoring a vital part of today's scientific picture. So, since the people I have just been considering are certainly vocal, for balance I have included them.

As it happens, although vocal, I do not really see these critics as being a very large segment of the scientific community, nor do I personally find the charges they bring, or the solutions they offer, that compelling. I say this without implying that nothing of value has been or is being said, and without denying that today we do face serious environmental and ecological problems. *Naturphilosophie* has always had a bad reputation, but clearly the men at the time did seize on matters of importance – the isomorphisms (what we now call homologies) between organisms, for instance. Like most people, however, I think Darwin put his finger on the causes of homology, and there is no need to break from the Darwinian machine-metaphor picture to explain things. The same is true today, despite Stephen Jay Gould's enthusiasm for things German. "Order for free" likewise strikes me as vastly overblown – no one ever denied that laws can produce complex and wonderful results, but whether this is the end of the matter is another issue entirely. In the case of phyllotaxis, for instance, conventional evolutionists have shown that even given the mathematical structures, natural selection still has a major role to play (Niklas 1988; Reeve and Sherman 1993). I suspect that some enthusiasts have spent too much time in front of computer screens and not enough looking at the real world.

Likewise, by the time Gaia has been cleaned up with the daisy-world picture, we have at most a very weak organic metaphor, and a fairly conventional mechanistic hypothesis is doing the heavy lifting. I should add that critics have argued that the real world is very much unlike the daisy world (Kirchner 2002). Finally, while one certainly has sympathy for ecofeminists in much that they write and bemoan, there is much debate about the actual position taken by people like Bacon – most scholars of the period deny strenuously that he was the chauvinist that he is portrayed to be (Vickers 2008). I confess that my soul blanches when I am told that I must adopt a position that does not find the eradication of smallpox to be a good thing, and as for the alternatives, I am reminded of what Mahatma Gandhi once said: "My friends tell me that it costs them a great deal to keep me in poverty." Fancy little homesteads with trout streams and views of snowcapped mountains, even (or rather especially) those that practice organic farming throughout, tend not to be the future homes of inner-city single moms living on welfare and food stamps. They are rather the weekend retreats of trial lawyers and other significant contributors to present-day culture. One thinks of Prince Charles and his enthusiasm for the organic, something comfortably backed by a massive income from the Duchy of Cornwall.

Experience suggests that what I and other critics of the organic model have to say will disturb its supporters not one bit. So let me at least hope that, although the model is not one that I am inclined to embrace, I have shown its richness and that it makes good sense as an overall world picture. By this I mean that if you think of the earth as an organism, then you can start to ask important questions and find strong answers within the paradigm. You do have a set of values that I suspect most people, within or without the model, do find sensible and in many respects compelling. To look to the well-being of the planet is something that should be important to each and every one of us. The metaphor-inspired picture may not be fully adequate, but one can respect someone who embraces it. It is rather like the attitude of a socialist toward a mainstream conservative. One may think that overall it is an inadequate philosophy, but one can respect someone who takes seriously traditions and standards and so forth.

BIOLOGICAL LIMITS TO KNOWLEDGE?

As with science done under the machine model, science done under the organic model leaves questions unasked and answers not given. In a way, we knew that before we started. Organicism is no less based on a metaphor than is mechanism. It was therefore bound to have its limits. Not to mention limits that organicism has that mechanism does not have – about the nature of life, for instance. As one final point before we move on, however, I do want quickly to link up the conclusions of this and the previous chapter with a pertinent, related line of argument. I am not sure how one would offer a transcendental argument that all science has to be metaphorical, although if one is going to have science exhibiting such epistemic virtues as predictive fertility, I am not sure how else it would be done. That suggests that all science is going to have unanswered questions. These need not be the same unanswered questions, of course, although our look at science done under two basic metaphors suggests that there may be some questions that science will never answer.

At this point, let us ask another question: "Why should we expect middle-range primates, with adaptations for getting out of the trees and onto the plains, to have the ability to peer into the ultimate mysteries of reality?" The wonder is that we can go as far as we can, not that we cannot go all the way. The eminent population geneticist J. B. S. Haldane mused about this: "My own suspicion is that the universe is not only queerer than we suppose, but queerer than we *can* suppose" (Haldane 1927, 286). Recently, Richard Dawkins has raised the same point: "Modern physics teaches us that there is more to truth than meets the eye; or than meets the all too limited human mind, evolved as it was to cope with medium-sized objects moving at medium speeds through medium distances in Africa" (Dawkins 2003, 19). I suspect that the Victorians would have been pretty dismissive of worries such as these. Notoriously, they were worried that science was practically exhausted and that there would be no new good problems to solve. Today, we are much less confident. For a start, Gödel's (already-mentioned) theorem in mathematics, or rather about mathematics, is deeply confidence-destroying. We now know that any reasonably interesting, consistent axiom system, especially one that deals with the natural numbers, is going to have true statements that cannot be proven within the system, no matter how much fiddling around you do with the system. This has led to huge amounts of discussion, including discussion about whether the theorem shows that the human mind cannot be a machine, because we can see the truth of the claims unprovable within the system – something we could not do if we were simply computer-based, calculating machines.

Not all of this discussion has been entirely edifying, as witness one line of thought responding to suggestions that humans are inconsistent thinkers. "Certainly women are, and politicians; and even male non-politicians contradict themselves sometimes, and a single inconsistency is enough to make a system inconsistent" (Lucas 1961, 120–1). But the point is not to explore this line here – if anything, it emphasizes how much we can know even where computations within a system might draw up short – but simply to show that complete answers to everything seem no longer quite as assured as they might once have. And quantum mechanics obviously backs up this point. Note that I am not now trying to say that quantum mechanics is mysterious in some bad way – as a prediction-generating apparatus, it is without equal – but it does obviously raise questions about the limits of human understanding. The most typical interpretation is to say that we simply cannot inquire into certain questions about the consistency of certain wave and particle effects. Even if one thinks that there is some answer, we cannot find it. Or, even worse, there is no answer that we could understand. We seem to have reached an outer point of what we can know. This is not a "science stopper" in the sense of an appeal to miracles – a common (and justified) complaint about those like the supporters of the Intelligent Design movement who want to give up looking for natural causes and turn to God – but it is a recognition that there are areas where we cannot go. (And recognizing this frees us to get on with the work that we can do.)

The problem, of course, is not so much recognizing that there will be limits but recognizing where and when we have reached those limits. Perhaps quantum mechanics is one place where we can be reasonably sure, but what about other places? Can there be general proofs about limits, or must we treat each case on its own merits (or demerits), and even then, how can we be sure? Recently, one school of thought about the body-mind problem has started to suggest that here is one place where there might not be solutions. Pinching the name of a rock group, they call themselves the New Mysterians. The British-born philosopher Colin McGinn (2000) is a leading spokesperson. He is adamant that here we might run out of intellectual steam. "It is easier not to know than to know. That truism has long had its philosophical counterpart in rueful admissions that there are nontrivial limits on what human beings can come to grasp. The human epistemic system has a specific structure and mode of operation, and there may well be realities beyond its powers of coverage" (McGinn 1997, 106). The body-mind connection may be one such reality.

McGinn wants to go beyond what we have offered so far, namely, the bare fact that every solution to the problem hitherto proposed is totally inadequate – the fact that Cartesian dualism leaves a gap that cannot be crossed and that materialism simply ignores the most important facts of consciousness. McGinn offers an argument suggesting that the problem can never be solved. He makes much of the fact that we are animals that evolved in space, and that space is important in our most basic thinking. Indeed, it could be that the only way in which we individuate objects is essentially spatial and hence that our very thinking about subjects and objects is a function of our biology in this respect. "We are, cognitively speaking as well as physically, spatial beings par excellence: our entire conceptual scheme is shot through with spatial notions, these providing the skeleton of our thought in general. Experience itself, the underpinning of thought, is spatial to its core. The world as we find it – the human world – is a preeminently spatial world" (ibid.). However, think of how distinctive is our notion of space. We are basically Euclidean beings, with the earth at the center of our existence. Again and again science has forced us to reject the commonsensical way of looking at things, until today we have all of the pressures to revise owing to relativity theory and quantum mechanics and the even newer theories about strings and so forth. "Our folk theory of space has been regularly hung out to dry. From the point of view of the divine physicist, space must look to be a very different creature from that presented to the visuo–motor system of human beings" (p. 105).

Could it be, therefore, that mind, an entity that puts such pressures on normal thinking about space, is yet one more step down the road of revising spatial thinking? We need a whole new theory or revision of what we have, but that will wreck what we have left of our basic thinking modes, which (remember) are deeply spatially engrained. And so we simply will not be able to think about things in the new way because our old ways are too fixed and limited. "Our conceptual lens is optically out of focus, skewed and myopic, with too much space in the field of view. We can form thoughts about consciousness states, but we cannot articulate the natural constitution of what we are thinking about. It is the spatial bias of our thinking that stands in our way (along perhaps with other impediments). And without a more adequate articulation of consciousness we are not going to be in the position to come up with the unifying theory that must like consciousness to the world of matter in space. We are not going to discover what space must be like *such that* consciousness can have its origin in that sphere" (p. 107). We cannot think properly about space. Thinking properly about space is essential to solving the body-mind problem. Hence, we cannot – and never will – solve the body-mind problem.

As you might expect, others in the field have little but scorn for McGinn's arguments. I will leave you to fill in the kinds of responses to be expected from people like Dennett and the Churchlands. The point to be made here is that whether you agree with McGinn or not, he is right in believing that we do think in ways dictated by our biology, and there is no very good reason to conclude that these ways are necessarily ones that guarantee a path to the understanding of the whole of absolute reality. We expect to run out of steam, the very conclusion reached by our study of the basic metaphors of science, ancient and modern.

SEVEN

GOD

Understanding is the reward of faith. Therefore seek not to understand that you may believe, but believe that you may understand.
Augustine 1873, 29.6.

Finally, we turn to religion. More precisely, we turn to the Christian religion, and even more precisely, we turn to the Christian religion of the West. Let us start with the Christian's basic statement of faith, the Apostles' Creed. I use the Anglican version, from the *Book of Common Prayer*:

> I believe in God the Father Almighty, Maker of heaven and earth. And in Jesus Christ his only Son our Lord; who was conceived by the Holy Ghost, born of the Virgin Mary, suffered under Pontius Pilate, was crucified, dead, and buried; he descended into hell; the third day he rose again from the dead; he ascended into heaven, and sitteth on the right hand of God the Father Almighty; from thence he shall come to judge the quick and the dead.
>
> I believe in the Holy Ghost; the holy catholic Church; the communion of saints; the forgiveness of sins; the resurrection of the body; and the life everlasting. AMEN.

Now, we have to whittle down the discussion somewhat, so I am going to brush past a great deal of Christian theology, especially

181

much that has to do with specific notions like the Trinity, important though they are. I am also going to ignore entirely much that makes Christianity vital and important to its practitioners – the churches, the music, the rituals, the fellowship, and all of that. Taking out the lofty spire at Salisbury and the glorious windows at Chartres, the agony of suffering that we hear in the Bach passions and the lusty singing of a Charles Wesley hymn, the dignified High Mass at Easter and the simple silence of a Quaker meeting, the eloquent and challenging sermons of Martin Luther King and the daringly innovative poetry of Gerard Manley Hopkins – that is to remove so much that gives Christianity its depth and vitality. But this must be. With an eye to the discussion of the previous chapters, I want to pick out four items or claims that are central to Christian belief – four items that the Christian takes on faith. If you do not believe in these, then you should not call yourself a Christian. First, that there is a God who is creator, "maker of heaven and earth." Second, we humans have duties, moral tasks here on earth, in the execution of which we are going to be judged. Hence, God stands behind morality. Third, Jesus Christ came to earth and suffered because we humans are special, we are worth the effort by God. The usual way of expressing this is to say that we are "made in the image of God." We have "souls." Fourth and finally, there is the promise of "life everlasting." We can go to heaven, whatever that means.

Let me spell out carefully what I see as the task in this and the next chapter. It is not to defend Christianity as a true or compelling belief system. I take it that you can enter these chapters as an agnostic or an atheist and depart in the same frame of mind. I do not want to dissuade people from Christianity, nor do I want to convince them of it. I want to explain in a fair manner what is meant by Christianity in the terms of the four points introduced in the last paragraph. I want also to show that you can hold these, if you so wish, in the light of modern science – if you prefer, in the face of modern science. In other words, the Christian's claims are not refuted by modern science – or indeed threatened or made less probable by modern science. In order to do this, I shall rely on the work of previous chapters, showing that there are certain areas that modern science not only does not answer but, as it is at the moment, does not even speak to. I am not saying that it could never speak to the areas – things like the previous century's work on the notion of life make me unwilling to make absolute statements – but I do not think that it does speak to them at the moment. It is not a question of trying and failing. It is more one of not being in the conversation. I shall argue that the Christian's claims fall within these areas, and so the Christian can, legitimately, try to speak to them.

I warn that that means that the Christian is not and cannot be offering a science-like answer. Certainly, the Christian should not be offering such an answer. The noted Christian (Calvinist) philosopher Alvin Plantinga (1999) comments about naturalistic accounts of the nature of mind: "A theist may be able to learn a good bit from this; but fundamentally he will ask different questions and look for answers in a quite different direction" (p. 19). Generalizing, this is precisely my claim. Although we must strive to understand science and its triumphs, questions such as that about ultimate origins are simply not attempted by science as we know it. My claim is also – and this is very important – that these questions are genuine questions. Hence, I argue that it is open for others to attempt answers to these questions. However, these must be answers of a different type in some way: nonscientific answers. For this reason, it is not fair to criticize the religious person for not offering a science-like answer. As it happens, the Christian claims to be giving a faith-based answer, one that comes from a different source than the reason and empirical experience (through the senses) that yields science. You may think that if science is silent, we should also be silent. That is your opinion, and others have the right to disagree. This does not preclude arguing against the Christian on nonscientific grounds. That is another matter.

FAITH

Getting ready to consider what Christianity says about God and about morality, souls, and eternity, we should say a little more about the all-important question of faith. How do Christians regard it, and how do they see the relationship between faith and other forms of belief support, especially reason? For Roman Catholics, faith is a form of intuition or discernment, something open to all human beings, a gift of God, a guide that leads us to an understanding of the truths of the Christian religion. In a sense, therefore, faith is akin to reason in the sense that it leads to truths of the same logical type.

A good starting point is *Fides et Ratio*, a fairly recent encyclical (letter to bishops on some point of doctrine) by Pope John Paul II:

> 9. The First Vatican Council teaches...that the truth attained by philosophy and the truth of Revelation are neither identical nor mutually exclusive: "There exists a twofold order of knowledge, distinct not only as regards their source, but also as regards their object. With regard to the source, because we know in one by natural reason, in the other by divine faith. With regard to the object, because besides those things which natural reason can attain, there are proposed for our belief mysteries hidden in God which, unless they are divinely revealed, cannot be known." Based upon God's testimony and enjoying the supernatural assistance of grace, faith is of an order other than philosophical knowledge which depends upon sense perception and experience and which advances by the light of the intellect alone. Philosophy and the sciences function within the order of natural reason; while faith, enlightened and guided by the Spirit, recognizes in the message of salvation the "fullness of grace and truth" (cf. *Jn* 1:14) which God has willed to reveal in history and definitively through his Son, Jesus Christ (cf. *1 Jn* 5:9; *Jn* 5:31–32). [The quotation is from the Dogmatic Constitution on the Catholic Faith, *Dei Filius*, IV: DS 3015.]

We get the content of faith from scripture and through the tradition of the church and the dicta of its leaders. It should be emphasized that although faith is a gift, we ourselves must make some effort. "Faith is said first to be an obedient response to God. This implies that God be acknowledged in his divinity, transcendence and supreme freedom" (14). And: "By the authority of his absolute transcendence, God who makes himself known is also the source of the credibility of what he reveals. By faith, men and women give their *assent* to this divine testimony. This means that they acknowledge fully and integrally the truth of what is revealed because it is God himself who is the guarantor of that truth" (79). Summing up, in the words of Saint Thomas: "To believe is an act of the intellect assenting to divine truth through grace so that the act rests on a free decision directed toward God" (ST II-II q. 2 a. 9).

There is significant overlap between Protestants and Catholics on faith, especially inasmuch as the great Reformers – Luther and Calvin – looked back, as do Catholics, to Saint Augustine. Calvin, for instance, states explicitly that God has implanted in us a natural tendency to believe in and understand the nature of God.

'There is within the human mind, and indeed by natural instinct, an awareness of divinity.' This we take to be beyond controversy. To prevent anyone from taking refuge in the pretence of ignorance, God himself has implanted in all men a certain understanding of his divine majesty. Ever renewing its memory, he repeatedly sheds fresh drops. Since, therefore, men one and all perceive that there is a God and that he is their Maker, they are condemned by their own testimony because they have failed to honor him and to consecrate their lives to his will. (*The Knowledge of God Has Been Naturally Implanted in the Minds of Men*, Calvin 1536, Book 1, Chapter 3, section 1)

Calvin worries a bit about primitive people. However, he reassures himself and his readers that "no nation so barbarous, no people so savage, that they have not a deep-seated conviction that there is a God." All humans have this sense or at least this capacity to come to realization of God. We are dealing with innate ideas here (the kind that John Locke was to deny in the next century): "that *there is some God*, is naturally inborn in all, and is fixed deep within, as it were in the very marrow.... From this we conclude *that it is not a doctrine that must first be learned in school*, but one of which each of us is master from his mother's womb and which nature itself permits no one to forget" (section 3). Calvin adds that although we have this natural tendency, it is obscured by our sinful nature. Hence, the existence of unbelief and doubt.

I should note that Plantinga (1983) argues that since we have this direct awareness or perception of God, it is properly in the domain of reason rather than faith, along with other basic claims of reason like the existence of other minds. This is not a matter of proof, as with natural theological claims (of which more in the next chapter), but of simply being at one with other things we include in reason. Faith claims are specific claims within the Christian religion, like the relationship of Jesus to God. But Plantinga does admit that others, nonbelievers especially, might find his thinking idiosyncratic, and that is certainly my position here. Importantly, whatever you call it, a direct awareness of God is not part of the corpus of modern science and must, if possible, be reconciled with that corpus.

Despite overlap on the nature and meaning of faith, we do find differences between Catholics and Protestants, starting with the fact that the latter rely only on scripture for faith beliefs, whereas the former turn also to tradition and the words of the leaders of the church.

(This difference is justified by the Catholics' believing that, insofar as Peter is the rock on which the church is founded, Jesus was giving a guarantee that Peter and his successors would get direct direction from God.) Perhaps more important is the Protestant feeling that faith is less a method of acquiring knowledge and more a dynamic acceptance of Jesus as savior, through whom is made possible eternal salvation – through whom one is "justified." The contemporary British philosopher of science and religion Richard Swinburne (2005) uses the word "trust." This aspect of faith is brought out well by Luther's coworker and supporter Philip Melanchthon. "Faith is not merely a knowledge of historical events but is a confidence in God and in the fulfillment of his promises. . . . We should understand the word faith in the scriptures to mean confidence in God, assurance that God is gracious to us, and not merely such a knowledge of historical events as the devil also possesses" (*Augsburg Confession*, art. 20). The great twentieth-century Protestant theologian Karl Barth speaks in similar terms: "Faith is the total positive relationship of man to the God who gives Himself to be known in His Word. It is man's act of turning to God, of opening up his life to Him and of surrendering to Him. It is the Yes which he pronounces in his heart when confronted by this God, because he knows himself to be bound and fully bound. It is the obligation in which, before God, and in the light of the clarity that God is God and that He is his God, he knows and explains himself as belonging to God" (Barth 1957, 12). Yet what is important here is that ultimately, for all Christians – Catholic and Protestant – these faith claims are also knowledge claims. Even Barth, who loathed all attempts to get at God through reason, allowed this: "What is certain is that faith must also be described as knowledge and can also be described thus in its totality" (p. 13). It is more a matter of how that knowledge of revelation can be obtained.

If faith is indeed knowledge, then one important point follows immediately. We must eschew "fideism," meaning an attitude that pits faith against reason – Tertullian (ca. 160–220 CE) notoriously (although possibly ironically) claimed that the resurrection is to be believed because it is "absurd," it is "impossible" (Roberts and Donaldson 1989, 525). Traditional Christian thinking is that faith cannot be unreasonable, in the sense of illogical. It might push you beyond reason, certainly beyond empirical evidence, but it cannot make you believe the contradictory. This is a point that will need more discussion later, but for now we are justified in making sure that the Christian's claims make at least some kind of sense. Otherwise it can hardly qualify as knowledge. This point was stressed by that eminently sensible philosopher John Locke:

> Indeed, if anything shall be thought revelation which is contrary to the plain principles of reason, and the evident knowledge the mind has of its own clear and distinct ideas; there reason must be hearkened to, as a matter within its province. Since a man can never have so certain a knowledge, that a proposition which contradicts the clear principles and evidence of his own knowledge was divinely revealed, or that he understands the words rightly wherein it is delivered, as he has that the contrary is true, and so is bound to consider and judge of it as a matter of reason, and not swallow it, without examination, as a matter of faith. (Locke 1689, 2, 424)

On this note, let us now turn to look at major planks of Christian belief. I shall not argue that you do not have to believe these things. That seems to me to be obvious. My inquiry is about those who do believe these things. Christians have faith. They believe that they have been told certain truths by God. What follows?

NECESSARY BEING

Let us plunge right in with the first claim:

> I believe in God the Father Almighty, Maker of heaven and earth.

This is a claim based on the Old Testament: "In the beginning God created the heaven and the earth. And the earth was without form, and void; and darkness was upon the face of the deep. And the Spirit of God moved upon the face of the waters" (Genesis 1:1–2). It relies likewise on words in the New Testament: "In the beginning was the Word, and the Word was with God, and the Word was God. The same was in the beginning with God. All things were made by him; and without him was not any thing made that was made" (John 1:1–3).

This claim is clearly intended to speak to one of the major questions we have seen to be unanswered by science: Why is there something rather than nothing? Does it do so, and does it do so in a way that does not infringe on the domain of science? In trying to answer this question, remember that I am going to do Christians the courtesy of assuming that we are not now dealing with absolutely literal readings of the Bible. I am assuming that we have moved beyond picturing God as He is depicted in Michelangelo's Creation of Adam, that is, as someone who looks a little bit like Charlton Heston, dressed in a bedsheet. It is still legitimate and necessary to ask something about the nature of God. As just noted, even though the Christian God is not a scientific concept, our understanding cannot violate the rules of reasoning (or the findings of the senses). We would be beyond knowledge as we know it, and we would be violating things that science does hold dear. Here, therefore, we start to grapple with somewhat philosophical issues, meshing with points made in an earlier chapter. God cannot be a contingent being like us. If he were, then we would get caught up in sophomoric questions like "God caused the world, but what caused God?" (When I say that such questions are sophomoric, I do not mean that they are unimportant questions. I mean it is sophomoric to think that you thought them up and that no one previously has tried to answer them.) God has to be a necessary being. Or let me phrase this more carefully, because I do not want to assert even implicitly that God exists. If God exists, then God exists necessarily. One immediate consequence of this is that God (assuming He exists) is not something in time. He is not just very old and likely to go on living for some time yet. He is eternal. He is like the truths of mathematics: $2 + 2 = 4$ did not start to be true, and it never will cease to be true. It is not like my dog Toby with a beginning, a middle, and an end. Saint Augustine is quite clear on this. This is part of God's being.

> Thou precedest all things past, by the sublimity of an ever-present eternity; and surpassest all future because they are future, and when they come, they shall be past; but Thou art the Same, and Thy years fail not. Thy years neither come nor go; whereas ours both come and go, that they all may come. Thy years stand together, because they do stand; nor are departing thrust out by coming years, for they pass not away; but ours shall all be, when they shall no more be. Thy years are one day; and Thy day is not daily, but Today, seeing Thy Today gives not place unto to-morrow, for neither doth it replace yesterday. (Augustine 1998, 11.13.16)

For Calvinists particularly, given their obsession with God's absolute sovereignty, His freedom seems to be His most important feature: "the prerogative of divine sovereignty." Unsurprisingly, therefore, for Barth, God's very being is all bound up with just this: "freedom in its positive and proper qualities means to be grounded in one's own being, to be determined and moved by oneself. This is the freedom of the divine life and love. In this positive freedom of His, God is also unlimited, unrestricted and unconditioned from without. He is the free Creator, the free Reconciler, the free Redeemer" (Barth 1957, 301). God's eternity follows from this sense of freedom, of God's being grounded in his own being.

> God's eternity, like His unity and constancy, is a quality of His freedom. It is the sovereignty and majesty of His love in so far as this has and is itself pure duration. The being is eternal in whose duration beginning, succession, and end are not three but one, not separate as a first, a second and a third occasion, but one simultaneous occasion as beginning, middle and end. Eternity is the simultaneity of beginning, middle and end, and to that extent it is pure duration. Eternity is God in the sense in which in Himself and in all things God is simultaneous, i.e., beginning and middle as well as end, without separation, distance or contradiction. (Barth 1957, 608)

This is not a matter of being everlasting, in the sense of being in time. Time is God's creation, but eternity is something more. "Time is distinguished from eternity by the fact that in it beginning, middle and end are distinct and even opposed as past, present and future. Eternity is just the duration which is lacking to time, as can be seen clearly at the middle point of time, in the temporal present and in its relationship to the past and the future. Eternity has and is the duration which is lacking to time. It has and is simultaneity" (ibid.).

The problem for the critic is that beings are never necessary. Their existence is always contingent. I have spoken to this in an earlier chapter (Chapter 5), but do not rely on me. Remember David Hume. We have here a difference with the truths of mathematics. It is impossible, for example, for the cube root of 64 to be equal to 5. In some sense, it is inconceivable. "But that CAESAR, or the angel GABRIEL, or any being never existed, may be a false proposition, but still is perfectly conceivable, and implies no contradiction" (Hume 1777, section xii, part 3, para. 28). The existence, therefore, of any being can be proved only by arguments from its cause or its effect; and these arguments are founded entirely on experience. There are many modern thinkers willing to follow in Hume's path. The philosopher

J. N. Findlay is one, arguing that God is not worth worshipping if He is not necessary and that this is simply not a possibility:

> [T]he Divine Existence could only be a necessary matter if we had made up our minds to speak theistically *whatever the empirical circumstances might turn out to be....* The religious frame of mind seems to be, in fact, in a quandary; it seems invincibly determined both to have that inescapable character which can, on modern views, only be found where truth reflects an arbitrary convention, and also the character of 'making a real difference' which is only possible where truth doesn't have this merely linguistic basis. If God is to satisfy religious claims and needs, He must be a being in every way inescapable, One whose existence and whose possession of certain excellencies we cannot possibly conceive away. And modern views make it self – evidently absurd (if they don't make it ungrammatical) to speak of such a Being and attribute existence to Him. (Findlay 1948, 182)

What is to be said in reply? It is certainly the case that there are those who want to claim that God's existence is in some sense logically necessary. Anyone who subscribes to the validity of the so-called ontological argument, which derives God's existence from his very nature, thinks along these lines. Saint Anselm's version shows this clearly: God is that than which none greater can be conceived. Existence is part of this. Hence, it is contradictory to think that God does not exist. However, this argument is certainly not part of the mainstream of Christian thought. There is nothing in the Bible suggesting that God is logically necessary. In both the Old and the New Testaments it is true that, in some sense, the awareness of God is so strong that His nonexistence is unthinkable – Job in his trials never denies God's existence, nor do his comforters – but this is more as a matter of fact than of logical necessity. After all, Anselm's ontological proof starts with a quotation from Psalms: "The fool hath said in his heart, There is no God" (53:1). This is not a sentiment the psalmist receives favorably, but neither is it an occasion to launch into a proto-form of Russell's theory of descriptions.

Anselm himself distinguishes a kind of logical necessity (to which his argument seems to point) from the more factual, which does seem to have biblical warrant. It is more a case of God being cause of Himself. "The supreme Substance, then, does not exist through any efficient agent, and does not derive existence from any matter, and was not aided by being brought into existence by any external causes. Nevertheless, it by no means exists through nothing, or derives existence from nothing; since, through itself and from itself, it is whatever it is" (Hick 1961, 357). This is a forerunner of the kind of thinking we find in Saint Thomas. He denied the validity of the ontological argument, but nevertheless thought of God as necessary. A kind of factual necessity, where God is thought of as responsible for His own existence. He is cause without cause. God is *a se* (hence has the property of "aseity"), being of Himself.

Aquinas offers several arguments or points of clarification on this. The first is that God is something wherein essence and existence come together.

> First, whatever a thing has besides its essence must be caused either by the constituent principles of that essence (like a property that necessarily accompanies the species – as the faculty of laughing is proper to a man – and is caused by the constituent principles of the species), or by some exterior agent – as heat is caused in water by fire. Therefore, if the existence of a thing differs from its essence, this existence must be caused either by some exterior agent or by its essential principles. Now it is impossible for a thing's existence to be caused by its essential constituent principles, for nothing can be the sufficient cause of its own existence, if its existence is caused. Therefore that thing, whose existence differs from its essence, must have its existence caused by another. But this cannot be true of God; because we call God the first efficient cause. Therefore it is impossible that in God His existence should differ from His essence.
> (Aquinas 1952, I q. 3 a. 4)

The second point is based on the notions of potentiality and actuality. God has no potentiality; He is purely actual. Hence He exists out of His very essence, and this in some way makes Him necessary. The third point picks up on participation and makes much the same point as the second. All of this is clearly bound up with an anticipation of Barth's claim about God being totally free. His being comes from Himself, and hence He is dependent on no one and no thing for His nature or His actions.

It is, to be candid, not easy (at least for the outsider) to see arguments here. It is perhaps better to regard Aquinas as offering a clarification of what is meant by aseity. So we are not getting a knockdown case for the truth of the Christian God. But perhaps that is not what we are after anyway. The question here is not whether any of this is true but whether it is in any sense possible. Logically necessary existence does raise problems, but that is not the claim here. The question is whether some kind of factually necessary existence is possible. It is not a spatio-temporal existence, a physical existence, but we know that. God's existence is not like ours. I suspect that if we want to think about these things positively, then, as I suggested in an earlier chapter, we should turn to Plato. He thought that the objects of mathematics have real existence (not of this world), very much like (although not the same as) the Forms, which, it will be remembered, included the Form of the Good, which was responsible for the being and welfare of everything else. Platonism in mathematics would not prove the existence of the Christian God, but it might at least make Him possible.

But can we make this move? Well, we can certainly make it, but whether it does any good is the big question. First, you have got to agree that something like π really does exist, and then, second, you have got to argue that this is relevant to the God question. As I said earlier, I find myself surprisingly comfortable with the first idea. I think the Euler identity $-e^{\pi i} + 1 = 0-$ is one of the most beautiful things I have ever seen. Something with the perfection of a Vermeer interior. And unlike the painting, it is not something in the creation of which humans had a hand. I am inclined to think that mathematical abilities are rooted in our biological past. Those proto-humans who thought in terms of simple arithmetic survived and reproduced, and those that did not did not. But (with Quine) I don't see why the powerful entities of mathematics should not be treated as theoretical terms or entities, much as we treat electrons in physics. The thing is that π is not a physical entity like an electron and does not depend on the contingent world for its existence. As I said before, I don't think any of this *has* to be true. It has to be plausible in the light of modern science. And this it is. The second move, from mathematics to God, is the stretch. But note that it is not a stretch that any of us are compelled to make. The question is whether the Christian can make it, if he or she so desires. It is a matter not of proving God from mathematics – although I suspect that some people, like John Polkinghorne (1994), are keen to do precisely this (Polkinghorne argues that the truths of mathematics could not be pure chance) – but of arguing that the necessary existence of mathematical objects is the thin end of a very large wedge – a wedge that opens the door enough for a reasonable belief in an eternal God. And I will simply say that, arguments to the contrary, notwithstanding, I think it does. We are not talking about Humean existence here, so critics like Dawkins are off base. We are talking about something else, the possibility of which modern science does not preclude.

(I am uncomfortable with going much further with this, but the thought does occur that if you accept mathematical Platonism and a Gödel-type power of intuition, then there might be an analogy here with the faith power of the Christian – something that likewise supposedly leads to understanding of nonphysical reality. I would, however, point to the disanalogy that whereas mathematicians can all agree on something like the Euler identity, religious believers often disagree about the objects of their faith.)

GOD'S PROPERTIES

Does any of this prove that God is the creator? I am not sure that it does. As opposed to Greek thought, which never, ever claimed that God (or the gods) created matter – for the Greeks, the stuff of the world was a given – it is central to Christian thinking that God created from nothing, *ex nihilo*. Together with the passages quoted earlier, the Bible harps continually on this theme. "By the word of the LORD were the heavens made; and all the host of them by the breath of his mouth" (Psalms 33:6). "Through faith we understand that the worlds were framed by the word of God, so that things which are seen were not made of things which do appear" (Hebrews 11:3). I don't see that we have addressed this issue at all. What we do seem to have done is to have moved God into the position where He could be the creator. He starts to be the sort of being that the Christian creator must necessarily be. That is no minor thing and is probably enough for our purposes, for we are not trying to prove the truth of Christianity but to make room for it in the face of science. Theologically, it is still open for the critic to go after the Christian, asking exactly how a necessary being might be a creator. Christians from Augustine on (and before) have wrestled with this very issue. But this is a somewhat different matter from faulting Christianity because it conflicts with modern science.

However, we cannot stop here, for the Christian wants to claim more of God than that He just exists, although note that already we are starting to bring in some properties such as His total freedom. God had the power to create. In fact, for the Christian, because of His total freedom, He has total power – He is omnipotent. He also has total knowledge – He is omniscient. And He did not create just for fun. He did it out of pure love. He wanted beings in His own image that He could cherish and care for, at the most extreme level. This means that He was prepared to intervene in His creation. "For us and for our salvation he came down from heaven, was incarnate of the Holy Spirit and the Virgin Mary and became truly human. For our sake he was crucified under Pontius Pilate; he suffered death and was buried." These are theological claims, and (you will be getting tired of my repeating this) they must be judged in this light. Christians believe these things on faith. But it is still incumbent on us to see how they measure up – or measure out – in the light of science. (In this paragraph, I have quoted the Nicene Creed (325 CE), which is somewhat later than the earlier version, the Apostles' Creed (circa 180 CE), quoted at the beginning of the chapter. It stresses that Jesus came to earth for our salvation.)

Part of the problem here, of course, is understanding precisely what one means by such terms as "all-knowing." The traditional Christian position, embedded in the thinking of Saint Thomas, has always been straightforward. We cannot know God directly, but we can know Him analogically. When we say that God is our Father, we do not mean that it was His sperm that fertilized our mother's eggs. We mean that God has fatherlike qualities – He is the ultimate creator, and He provides for us, and He cares for us. Just remember, this is always a bit imperfect: "It should nonetheless be kept in mind that Revelation remains charged with mystery. It is true that Jesus, with his entire life, revealed the countenance of the Father, for he came to teach the secret things of God. But our vision of the face of God is always fragmentary and impaired by the limits of our understanding" (John Paul II 1998, 12–13). It is precisely this point that is stressed by many Protestants. Barth was a prime example. He wanted to stress above all the otherness of God, and this makes Him rather distant, at least with respect to understanding.

> In the act of the knowledge of God, as in any other cognitive act, we are definitely active as the receivers of images and the creators of counter-images. Yet while this is true, it must definitely be contested that our receiving and creating owes its truth to any capacity of our own to be truly recipients and creators in relation to God. It is indeed our own viewing and receiving. But we ourselves have no capacity for fellowship with God. Between God and us there stands the hiddenness of God, in which He is far from

us and foreign to us except as He has of Himself ordained and created fellowship between Himself and us – and this does not happen in the actualising of our capacity, but in the miracle of His good-pleasure. Our viewing as such is certainly capable of receiving images of the divine. And our conceiving as such is certainly capable of creating idolatrous pictures. And both are projections of our own glory. But our viewing and conceiving are not at all capable of grasping God. (Barth 1957, 182)

However, as Plantinga (1980) rightly points out, you can only push this line of argument so far, or else it becomes unclear whether worship is at all possible or desirable. Why should one bow down before something completely unknown? "If our concepts do not apply to God, then he does not have such properties as *wisdom, being almighty* and *being the creator of the heavens and the earth.* Our concept of wisdom applies to a being if that being is wise; so a being to whom this concept did not apply would not be wise, whatever else it would be" (p. 22). You can generalize this point: "If, therefore, our concepts do not apply to God, then our concepts of being loving, almighty, wise, a creator or a Redeemer do not apply to him, in which case he is not loving, almighty, wise, a creator or Redeemer. He won't have any of the properties Christians ascribe to him" (ibid.).

Start running through the properties. All-loving. One can certainly glimpse without contradiction what it would mean for a being to be all-loving. Such a being would give unconditional love – freely, without hope of gain, all of the time, as much as could be given. Parents know what it is to give to their children. Teachers and professors know what it is to give to their students. Just keep multiplying this. Of course, what constitutes love in particular instances is often problematic – we all know parents (perhaps ourselves) who spoil their children with too much unconditional giving, but love never implied being soft or avoiding difficult decisions.

All-powerful. There are Christians today who want to avoid or modify this attribute. Process philosophers, those influenced by the philosophy of Alfred North Whitehead, think that God is part of the creative process along with us (Whitehead 1929; Cobb and Griffin 1976). They often invoke the notion of kenosis, that is, of God voluntarily giving up His powers, as when He let himself be crucified on the cross. ("Let this mind be in you, which was also in Christ Jesus: Who, being in the form of God, thought it not robbery to be equal with God: But made himself of no reputation, and took upon him the form of a servant, and was made in the likeness of men: And being found in fashion as a man, he humbled himself, and became obedient unto death, even the death of the cross" (Philippians 2:5–8). For the process thinkers, God can help to direct things here on earth, but He cannot force anything (Ruse 2008b). This is still very much a minority position. Most Christians, especially those in the Augustinian tradition (which very much includes Calvinists) do not accept these limits (voluntary or not) on God. But one must tread carefully. Does the traditionalist mean that there are no limits whatsoever on God's powers? Can one think coherently of a being who has total control and can do anything? Every first-year undergraduate knows that there are pitfalls here. Could God make a stone that He could not lift? Could God make $2 + 2 = 5$? Could God commit suicide? Generally, the answer has been that omnipotence does not mean being able to do the logically or mathematically impossible – and, even if God could commit suicide, He would never do so because this would go against His always wanting to do good, and killing a perfect being would hardly be a good act. Hard cheese on us too if God tops Himself.

Admittedly, not everyone has agreed on these points. Descartes notoriously thought that God could make mathematics do anything He wanted if He were so inclined. (This is known by philosophers as "universal possibilism.") As someone with inclinations toward Platonism in mathematics, I am inclined to think that God is bound by the laws of mathematics (and of logic) – or, if you prefer to put it another way, God's omnipotence is not constrained by the fact that he cannot do the logically or mathematically impossible. However, if you warm toward universal possibilism, you might still want God bound by logic, even if you argue that in an age of non-Euclidean geometries (which seemed very non-possible to people before the nineteenth century) God has more freedom of action when it comes to mathematics. One suspects that in the end, full-blooded universal possibilism verges on the incoherent – to us at least. "If we cannot understand 'infinite power,' we also cannot understand and hence cannot believe or know, the proposition that God's power is infinite" (Plantinga 1980, 116). Remember John Locke: "if anything shall be thought revelation which is contrary to the plain principles of reason,... there reason must be hearkened to, as a matter within its province." Certainly, if you want to be on-side with science, Christians are constrained in their thinking on these matters by the usual rules of reason. Even if you go so far as to think that God can do the logically impossible (even if He never would), you have to explain this to yourselves and to us in terms that do not imply the logically impossible. (In the next chapter, I will have more to say on claims that go beyond understanding as we have it or indeed as we could possibly have it.)

THE PROBLEM OF EVIL

Raise now the venerable problem of evil. If God is all-powerful, then He could prevent evil, and if He is all-loving, then He wants to prevent evil; and yet evil exists. Hence God cannot have both of these predicates usually ascribed to Him. Whether or not you think them adequate, the Christian has moves that avoid outright contradiction (Pike 1964; Hick 1978; Ruse 2001). With respect to moral evil – the evil that led to Auschwitz – the Christian (following Augustine) invokes human freedom. It is better that humans have freedom, even though it led to the gas chambers, than that we be simply automata. With respect to physical evil – the child dying of leukemia – the Christian (following Leibniz) argues that it is all a matter of possibilities. Since God is constrained by the impossible, if He wanted to maximize the good things, necessarily He had to create a world containing physical evil. The good from being warned outweighs the pain from burning. Gravity is a good, even though people fall to their deaths from heights. You may or may not find these responses adequate. Notoriously, Voltaire was scathingly funny in his *Candide* about the Leibnizian argument. That is not our question now. Agreeing that the moves would let the Christian avoid outright contradiction, the question now is: Does science exacerbate the problem of evil? If you hold to modern science, does this make the problem of evil even more intense, to the point that reasonable Christian belief is impossible?

Many scientists and fellow travelers think that this is so. In particular, evolutionary biology, with its focus on a bloody struggle for existence, the force behind natural selection – the main cause of evolutionary change – mocks the idea of the Christian God. Richard Dawkins, predictably, has been eloquent on the subject. The definitive text is a letter that Darwin wrote on May 22, 1860, to his friend Asa Gray, a Harvard botanist and a deeply committed evangelical Presbyterian.

> With respect to the theological view of the question; this is always painful to me. – I am bewildered. – I had no intention to write atheistically. But I own that I cannot see, as plainly as others do, & as I sh^d. wish to do, evidence of design & beneficence on all sides of us. There seems to me too much misery in the world. I cannot persuade myself that a beneficent & omnipotent God would have designedly created the Ichneumonidae with the express intention of their feeding within the living bodies of caterpillars, or that a cat should play with mice. Not believing this, I see no necessity in the belief that the eye was expressly designed. (Darwin 1985–, 8, 224)

Dawkins agrees wholeheartedly with this sentiment. Predators are designed for catching prey, prey for evading predators. If one wins, the other dies in the agony of skin-and-flesh-ripping fangs and claws. If the other wins, then the other dies in the agony of starvation. This is not just chance. This is the way that things are and must be. "If Nature were kind, she would at least make the minor concession of anesthetizing caterpillars before they are eaten alive from within. But Nature is neither kind nor unkind. She is neither against suffering nor for it. Nature is not interested one way or the other in suffering, unless it affects the survival of DNA. . . . The total amount of suffering per year in the natural world is beyond all decent contemplation." But the point, Dawkins stresses, is not that nature is intentionally vile. It is just that nature is indifferent. Remember: "DNA neither knows nor cares. DNA just is. And we dance to its music" (Dawkins 1995, 133).

Many others agree strongly with the sentiments of Darwin and Dawkins. As we know, the philosopher Philip Kitcher (2007) has recently recanted a lifelong willingness to see the possibility of Christian faith and practice. In terms appropriate to a revival tent, he has confessed his sins and now agrees that if one subscribes to Darwinian evolutionary thinking, then it is simply not possible to commit to the Christian religion – at least, it is not possible if one wants to retain some modicum of integrity and to deserve the attribute of reasonable being. Unfortunately, although perhaps predictably, one suspects that a lot of Christians feel the same way. The Reverend Keith Ward, an Anglican priest and sometime Regius Professor of Religion at Oxford University, seeks a warmer, friendlier form of evolutionary theory. We may not be able to get away from suffering and death, but at least we can tone them down a little: "On the newer, more holistic, picture, suffering and death are inevitable parts of a development that involves improvement through conflict and generation of the new. But suffering and death are not the predominating features of nature. They are rather necessary consequences or conditions of a process of emergent harmonisation which inevitably discards the old as it moves on to the new" (Ward 1996, 87).

Before we get bowled over, let us look at these worries in the light of the Christian responses to the problem of evil. The questions for us to ask are about how these responses fare when faced with modern science, specifically with Darwinian evolutionary theory. Start with the free will defense against moral evil. Evolutionary theory is of course something that, like the rest of science, supposes that the world runs in a lawlike fashion. We need not dwell long on this point because we have already spoken to the fear that law in some sense precludes free will. I have taken a compatibilist position, arguing that free will demands law rather than fights it; but, even if you are not a compatibilist, you are still going to agree that the world is lawlike and that this applies to living things, including humans. The whole of child rearing and education presupposes this. Why try to teach the multiplication tables to children unless you think that they are animals capable of learning them and of retaining them for more than five minutes? If you are a noncompatibilist, then you think that there is more to life than law, but we knew that already. The real question is whether Darwinian evolutionary biology in its own right in some sense makes the free will defense less compelling than otherwise. You might think so. A major cry about thirty years ago, repeated to this day, is that to apply Darwinian theory to human behavior is to suppose an illegitimate thesis about "genetic determinism." You are supposing that humans really are bound by laws, with their genes calling the shots, and that freedom is no more real than the hypnotized stage volunteer's illusion that he or she is free. Truly we are marionettes at the mercy of the controlling DNA (Lewontin et al. 1984).

In fact, as Daniel Dennett (1984) has pointed out, this simply is a travesty of the Darwinians' thinking on human nature. Ants are genetically determined. They are, to use a metaphor, entirely controlled by the tiny computers that they have for brains. They do not think. They just obey orders. And this is just fine for ants, but it would not be just fine for humans. The reason is simple. Ants have gone the route of having many offspring along with having little care or concern for individual well-being. The great thing about being genetically determined is that you do not have to waste effort on education and so forth. The really bad thing about genetic determination is that if something goes wrong, you do not have the ability to put things right. Dennett gives a lovely example. He tells of a wasp that brings food to its nest to provision its young. "The wasp's routine is to bring the paralyzed cricket to the burrow, leave it on the threshold, go inside to see that all is well, emerge, and then drag the cricket in. If the cricket is moved a few inches away while the wasp is inside making her preliminary inspection, the wasp, on emerging from the burrow, will bring the cricket back to the threshold, but not inside, and will then repeat the preparatory procedure of entering the burrow to see that everything is all right" (Dennett 1984, 11). This can go on and on indefinitely. "The wasp never thinks of pulling the cricket straight in. On one occasion this procedure was repeated forty times, always with the same result." This is the problem. A thousand ants are out foraging, and its starts to rain. A thousand ants are lost because the chemical (pheromone) trails that lead them home are washed away.

In the case of the ants, it does not matter much. There are hundreds of thousands more where the vanished ants came from. In the case of humans, it would matter very much. We have gone the route of having few children who demand much care. We cannot afford to lose even two or three every time it starts to rain. So how have we set about raising children in the face of adversities like rainstorms and predators and fellow humans? Here is where our brains are important. As Richard Dawkins says in *The Blind Watchmaker* (1986), we have big onboard computers, and when we come to challenges we can reason how to overcome them. This does not mean that we move from beneath the net of law, but that we have a dimension of freedom that the ants do not have. To continue with Dennett's examples, we are like the Mars Rover (a machine, take note). When it came to a rock, it did not stop, but had the computer-driven ability to reason how to get around the rock. Or to use an example of my own, ants are like cheap rockets that are aimed at the target, but if the target moves they miss it because they cannot change direction. Humans are like expensive rockets that have homing devices to track a moving target. Both rockets do what they do because of unbroken law, but the expensive rockets – aka humans – have a dimension of freedom that the inexpensive rockets – aka ants – do not. In other words, I suggest that far from Darwinian theory blowing holes in the free will defense, if anything it comes to its aid. (Technically, biologists say that humans are practising K selection, whereas the ants are practicing r selection. Of course, everyone thinks that there has been a feedback process in evolution. As we got better at raising children, we could have fewer, and having fewer put more selective pressure on being better able to raise children.)

What about the physical evil defense that God cannot be expected to do the impossible? Bad things are inevitable as good things are maximized. Here again matters are a little more complex, and more paradoxical, than you might expect. As Dennett came to the aid of the free will defense, so Dawkins comes to the aid of the Leibnizian defense. He argues that we could not have a functioning organic world without adaptation – designlike features – and that the only way in which you can get such features is through natural selection. The Intelligent Design theorists' supposition of guided mutations or variations is just not science. Lamarckism, the inheritance of acquired characteristics, is false. Macro-mutations or large variations leading to instant new features simply do not create designlike effects. The second law of thermodynamics kicks in here – things run down and not up, creatively. Chance leads to mistakes and not to triumphs of design. This really leaves only selection.

> My general point is that there is one limiting constraint upon all speculations about life in the universe. If a life-form displays adaptive complexity, it must possess an evolutionary mechanism capable of generating adaptive complexity. However diverse evolutionary mechanisms may be, if there is no other generalization that can be made about life all around the Universe, I am betting it will always be recognizable as Darwinian life. The Darwinian Law . . . may be as universal as the great laws of physics. (Dawkins 1983, 423)

For the Christian, the logic now is simple. God has created through unbroken law. There are fairly good theological reasons why He did things this way and not in one instantaneous flash of creative activity. If you are an Augustinian, thinking the thought of creation, the act of creation, and the product of creation are as one to God – because He stands outside time – then it is very natural to see God's creation as unfurling rather than pinned down to one moment. Augustine himself recognized this, because he spoke of God having created the seeds of life rather than life itself. But what are the consequences of this creation through law? Adaptation is necessary. The only way you can get adaptation through law is by natural selection. Natural selection demands a struggle for existence. A struggle for existence necessarily involves pain and suffering. The virtues of having animals, humans particularly, outweighs the pain and suffering. Hence, God could not have done other than He did, given His all-loving nature.

OMNISCIENCE AND HUMAN FREEDOM

Pushing aside universal possibilism, none of this compromises God's freedom. Which is just as well, for, as we have seen, part of God's very being is His freedom. Indeed, this freedom is such a great good that, as part of His love, God gave humans freedom also. There are immediate theological problems here, especially for the Calvinist, who stresses God's sovereignty and total power. How much freedom can humans be allowed? Notoriously, Calvin (and his great inspiration Augustine) embraced the doctrine of predestination – God foreordaining all that will happen, especially the choice of the saved and the damned. Whether this now permits any genuine sense of human freedom is a problem for the Christian theologian, not for us. Note that neither compatibilists nor incompatibilists like libertarians (understanding this term in the philosophical sense) are denying that we are free. Admittedly, there is a group with the somewhat ugly name of "hard incompatibilists" who do deny freedom; but even they, in the kind of paradox beloved of philosophers, nevertheless argue that you can have morality and a meaning to life – in a sense, the loss of freedom "may even be beneficial"! (Pereboom 2007, 85) There could be a temptation to think that for the Christian the only true kind of freedom is one that escapes the machine metaphor. Perhaps this is so. Nothing said in earlier chapters points to the futility of even trying to make this case. However, before rushing to judgment and concluding that this shows at least a potential point of stress with science, because (something I freely admit sways me) the success of modern science inclines (without necessarily being definitive) toward compatibilism, reflect that for the predestinarian there could be virtues in the compatibilist option – we have freedom, and yet God as creator of the laws knows (and determines) how

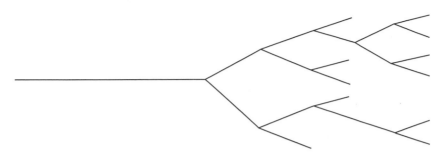

Figure 7.1. The Garden of Forking Paths.

things will come out. David Hume, raised a Scottish Calvinist, did not come out of nowhere.

Omniscience raises a related problem about human freedom. God is a bit like your mom. He knows exactly what you are up to. Because He is creator, he also knows about everything else. However, if God knows everything, He knows what each and every one of us will do before we do it. Which means, apparently, that there is truly not much freedom about what we do. We are free, and yet we are not free. Again, the compatibilist may feel a little smug, for in Dennett's felicitous phrase (in the title of his 1984 book), we still have all of the varieties of freedom "worth wanting." But what if you want a more radical form of freedom? What if you think freedom demands (in a phrase taken from the novelist Jorge Luis Borges) a "garden of forking paths" (Kane 2007) – many different possible outcomes? [Figure 7.1] One solution is to say that beliefs about the future are not really knowledge, but this seems unsatisfactory in an age of science. An astronomer really does *know* when an ellipse is about to occur. Another solution, favored by Augustine, is to say that knowledge of what someone is about to do does not negate that person's freedom.

> Unless I am mistaken, you would not directly compel a man to sin, though you knew beforehand that he was going to sin. Nor does your prescience in itself compel him to sin even though he was certainly going to sin, as we must assume if you have real prescience. So there is no contradiction here. Simply you know beforehand what another is going to do with his own free will. Similarly, God compels no man to sin though he sees beforehand those who are going to sin by their own free will. (Pike 1970, 76, quoting St. Augustine, *De Libero Arbitrio*, Book III)

The situation is rather like an examination. You know that Mary is going to do better than John, but how they perform is a function of their own abilities and study habits and so forth. Even here one might worry that God is too involved – who made Mary different from John? Hence, probably a stronger response is one that is certainly hinted at in Augustine, although developed most fully by Boethius. Here one appeals to another of God's attributes, namely that He is eternal – He stands outside of time. Therefore, for God there is no past, present, or future – He sees what happens in the exam now as he likewise sees the person studying for the exam earlier and getting the diploma afterward. His kind of knowledge about the future is not the same as our kind of knowledge about the future. Ours does cut down on choice; his does not.

> [T]he knowledge of God does not in any way vary: He does not know in different ways things which are not yet, things which are now, and things which are no longer. For he does not, as we do, look forward to what is to come, at what is present, or backward upon what is past. On the contrary, He views things in quite another fashion than we do, and in a way far and greatly different from our manner of thought. For His thought does not change as it passes from one thing to another, but beholds all things with absolute immutability. Of those things which occur temporally, the future, indeed, is not yet, the present is now, and the past is no longer; but all of these are comprehended by Him in His stable and eternal presence. Neither does He see in one way with the eye and in another with the mind, for He does not consist of mind and body. Nor does what He knows now differ from what He has always known and always will know; for those three things of time which we call past, present and future, though they affect our knowledge, do not change that of Him 'with Whom is no variableness, neither shadow of turning'. (Augustine 1998, 475, Book XI, Chapter 21)

What you gain on the roundabout you lose on the swings. If this solution does work, it seems to me that it does so only by making God that much more distant from, different from, us – that less easy to comprehend. However, this is not really my problem. It is enough to show that Christians have proposed solutions to these problems, solutions that do not obviously violate the claims of science.

MIRACLES

God created heaven and earth. For the Augustinian, God always had this thought of creation, and so one might suppose that, in a sense, the existence of God and the existence of matter are a package deal. Although Genesis gives the impression of a definite moment of creation, a point at which presumably time began, in a deeper sense you never get God without His creation. However, as always, beware of pantheism, which holds that the world is God or a part of Him. Christian theology is clear that God and the creation are separate, although it is also clear that with the creation completed, God's job is not done. He has to sustain the world, every moment of its existence. Without God, there would be nothing. This means that it is open to God to interfere in the world. And this is precisely what Christians believe that God did do. Most significantly, God intervened by sending Jesus to die for our salvation – a death followed by the Resurrection and the possibility of eternal salvation. Supposedly there were lots of other interventions – miracles – from events in the Old Testament like the parting of the waters to let the Israelites escape, through the various miracles of Jesus, to those performed by the disciples. For Catholics, especially, miracles are still-occurring phenomena.

But does not the belief in miracles fly in the face of science? What we mean by natural science is the attempt to explain the physical world, the world of experience, in terms of unbroken regularity, of law. It is true that back in the seventeenth century some, – Newton himself, prominently – were prepared to allow God some kind of intervening and adjusting role in the universe's running, but at least since the nineteenth century this has not been permitted. William Whewell, ardent though he was against evolution, admitted that his own appeal to the supernatural took him out of the realm of law. Of organisms he wrote: "when we inquire whence they came into this our world, geology is silent. The mystery of creation is not within the range of her legitimate territory; she says nothing, but she points upwards" (Whewell 1837, 3, 588). The argument and the conclusion are obvious. Science demands unbroken law. Christianity posits miracles, violations of law. Hence, science and Christianity are in conflict. Even though Christianity may speak to science's unanswered questions, science cannot accept the legitimacy of the Christian faith.

Traditionally, Christians have two responses to this charge (Mullin 2000). First, in the spirit of Augustine, one argues that miracles do not necessarily involve a breaking of natural law. It is more a matter of meaning than of peculiar happenings. Miracles are not against nature, *contra-naturam*, but beyond our knowledge of nature, *supra-naturam*, producing wonder. When Jesus turned the water into wine and when he fed the five thousand, it is much more probable that people were really moved by his presence and preachings to do that which they would not have done otherwise. The host at the wedding had not meant to bring out his good wine, and then, shamed by Jesus, he did just that. The multitude had no intention of sharing and then, moved by Jesus, they did give to those who had little or none. In fact, a case might be made for saying that this puts Jesus in a better light than if he actually were changing water into wine and multiplying loaves and fishes. To think literally is to change Jesus into some kind of fancy caterer.

In the case of healing, it is well known that charismatic people can have a huge influence on the ways in which people think and behave. Clearly Jesus had healing powers, but whether they were supernatural is another matter. For those arguing in this mode, even the Resurrection can be given a natural explanation (Williams 2000). Most probably Jesus did die, and his physical body rotted in the tomb. But the disciples felt his presence. By rights they should have been depressed and downtrodden. The very contrary was the case. They felt a surge of comfort and power. They knew that Jesus' death had not been in vain. He had conquered death in the only sense that was needed. They were thus emboldened to go forth and to preach the gospel. This was what happened and was all that needed to happen. That there may be a naturalistic explanation for all of this, something in terms of group psychology, is absolutely irrelevant. It is what you would expect. It was the meaning of it all that made for the miracle. (Rudolf Bultmann [1958] was the great twentieth-century spokesman for this position.)

The other approach, more in tune with Saint Thomas, is to agree that there were real violations of the laws of nature – done by divine agency beyond the order commonly observed in nature (*praeter ordinem communiter observatum in rebus*) – but to invoke the distinction between the order of nature and the order of grace. Normally, the world works by law, but our salvation required that God intervene in His creation – most importantly, by sending Jesus and allowing the events that led to the cross and then to the empty tomb. The Resurrection was truly a violation of the laws of physiology, and Jesus truly did come back to life after two days of death, but this was something that God allowed – enacted – because otherwise we humans would not have been saved from our sins. (Wolfhart Pannenberg [1968] speaks for this tradition.) Note that there is an important distinction here between the miracles of Christianity and the miracles of something like Intelligent Design theory. The latter is trying to do the work of science, arguing that unbroken law is not enough to furnish the living world with the complex organisms that it contains. The Christian miracles are not trying to do the work of science at all. They are about something entirely different, and they are simply laid across the world of science.

Care needs to be taken here, perhaps more care than some Christians are prepared to show. There are those who want to prove the Christian miracles empirically; Pannenberg believes that the stone was really moved and that the Resurrection is therefore a plausible event in history (Holwerda 1983). My suspicion is that David Hume (1777) was right and that the evidence against miracles is always more plausible than the evidence for them. Proving miracles gets you into the domain of science, and that is a sucker's game. Better simply to accept the miracles on faith and leave it at that. Also, one of the important things about miracles is that generally they are rare – they are miraculous! If too many miracles are called for, then the notion starts to get debased. It is true that some of the most common miracles are conceived in such a way that, even though God's specific action may be called for, they do not involve the breaking of natural law. The Catholic miracle of transubstantiation – where, in the Mass, the bread and the wine turn into the body and blood of Christ – is something not open to empirical check or refutation. Generally, however, where there are supposed violations of natural law, it is better that Christians not play the grace card too often. Or one wonders how serious they are about preserving the integrity of science.

MOVING ON

There is surely much more that one could say about the Christian notion of God. But, with major points now covered, the major thrust of the argument should now be obvious.

EIGHT

MORALITY, SOULS, ETERNITY, MYSTERY

Let us move now through the other items identified as key claims for Christians.

MORALITY

We know that science can tell us much about morality. For the sake of argument, let us take it now that biology can throw considerable light on these issues – moral behavior is ultimately an adaptation put in place by natural selection to make us functioning social beings. And it goes without saying that, in understanding morality, cultural investigations also have a major role to play. We in the Western world think that female circumcision is a cruel mutilation of women and girls. People in many parts of Africa think very differently. People in Canada and in Europe think that capital punishment is cruel and barbaric. People in the United States, especially in the South, think that capital punishment is morally obligated. There should be more of it. However, for all that science is important, we have seen also that ultimately science falls away. Thanks to the insights of David Hume, we know that no scientific justification can be given for moral beliefs. There is no metaethical foundation to be found in nature.

This conclusion follows immediately under the mechanical model and, if a little more slowly, also finally under the organic model.

You can at this point opt for some kind of ethical skepticism (about foundations). Moral beliefs and behaviors are adaptations, and there is an end to it. Others perhaps want to go different routes, feeling uncomfortable with denying that there is any justification at all. Presumably, then, your options are to think either that there are moral facts of some sort, or that there are not but that there is something more than psychology. If you think there are moral facts, then they have to be non-natural or you are back into justifying morality with science. If they are non-natural, then they have to be entities a bit like Plato's Form of the Good that we can intuit through a kind of moral sense and that are binding on us. Mathematics again comes to mind. We must be good just as we must accept that two plus two equals four. Sometimes we ignore one or both, but they are still binding. If you want morality without moral facts, then probably you are going to be drawn to some kind of Kantian solution, where morality is a condition of rational beings living together. In which case you are still going to be subject to the rules of applied mathematics – game theory and the like – not to mention having to explain how you get beyond the formal content of what you claim. As Kant himself recognized, you need more than just reciprocation. You need the form that the reciprocation can take – for example, "Love your neighbor as yourself." Indeed, Kant went so far as to say that we could be selfish and have a functioning society. "If such a way of thinking were a universal law of nature, certainly the human race could exist, and without doubt even better than in a state where everyone talks of sympathy and good will, or even exerts himself occasionally to practice them while, on the other hand, he cheats when he can and betrays or otherwise violates the rights of man" (Kant 1959, 41). It is because of this need (to fill out the content of human nature) that, when push comes to shove, I am not sure that the kind of Humean position that I favor is in these respects so very different from the Kantian position.

The important point is that with nontheological metaethics there comes a moment when the explanations have to stop. You have to accept that this is how existence is, whether this existence be purely natural or something more. If you are asking for something ultimate beyond that, you are out of luck. And it is no longer worth working harder at your science, because it is no longer science's business.

It is true that, especially with respect to ethical skepticism, many would disagree and would trot out the oft-made jibe that, without something ultimate, one is leaving the way open for unrestricted rape and pillage. But this is not much of an argument. Even the ethical skeptic can properly respond: "I am a human like everyone else, I am not a psychopath with a damaged or absent sense of right and wrong, so I want to do good like everyone else – as indeed I frequently fail like everyone else." The nonbeliever can ask why the lack of belief in a philosophical foundation for morality demands at once that one go against one's shared (with other human beings) psychological nature.

Non-natural facts, necessary conditions for social interaction, or not, the Christian wants to go another step (or two or three). "I agree with you about what you have said and claimed, but I still want more answers. I still want to know why things are the way they are morally and not some other way. I still want to know why it is really not right to take all of the cake for myself and not share. I know all of the biology and psychology, not to mention the philosophy, but why is greediness a sin?" It may be redundant, but I don't think it illicit to ask this question. I don't see that the Humean can rule it out of court. But can the Christian offer a coherent answer, within the Christian framework? Can the Christian add something, without messing things up scientifically? The beginning point has to be the belief that God is supremely good and has made us potentially good. Morality, therefore, is something that comes from God's nature, and since God is a free being, it is something bound up with God's will. God wants goodness, and God wants us to be good – where being good is what God wants of us. In other words, I suggest that the Christian is virtually bound to be committed to a form of what is known as the "divine command theory" of metaethics.

At once flags will go up. Several centuries before Jesus, in the *Euthyphro*, Plato (or, more likely, Socrates) put his finger on the weakness of this argument. Could God will the immoral? Is that which is good, good because it is the will of God; or is the will of God good, because it is good? If God wanted rape and pillage to be the rule and hence the good, would they be good? Surely not. In which case, it would seem that there is an authority even beyond God. However, the Christian – particularly the post-Aquinas Christian, that is to say, the Aristotelianized Christian – has a strong tool to avoid being caught in this trap (Quinn 1978): natural law theory. First, there is eternal law: "This rational guidance of created things on the part of God ... we can call the Eternal law." Then, we humans fit into this: "This participation in the Eternal law by rational creatures is called Natural law" (Aquinas 1952, 1a2ae, qu. 91, art. 1). The point is that God does not just decide things capriciously. Human morality has to fit with human nature, and this means (here is the Aristotelian component) with what is natural, with what is the proper functioning of our being, physical and mental. In other words, what is right is what is in tune with the intended final causes of our nature.

Take, by illustration, a matter of contemporary controversy: male homosexuality (Ruse 1988b). The traditional natural law theorist like Aquinas says that this is wrong because vaginas were made for penetration by penises and anuses were not. The intended use of the latter is defecation. Hence male-male buggery is wrong because the organs are not being used as they were intended.

> We have said that God exercises care over every person on the basis of what is good for him. Now, it is good for each person to attain his end, whereas it is bad for him to swerve away from his proper end. Now, this should be considered applicable to the parts, just as to the whole being; for instance, each and every one of his acts, should attain the proper end. Now, though the male semen is superfluous in regard to the preservation of the individual, it is nevertheless necessary in regard to the propagation of the species. Other superfluous things, such as excrement, urine, sweat, and such things, are not at all necessary; hence, their emission contributes to man's good. Now, this is not what is sought in the case of semen, but, rather, to emit it for the purpose of generation, to which purpose the sexual act is directed. (Aquinas 1975, 143:III, 22)

Homosexual acts are "unnatural vices," *vitae contra naturum*. In the words of Pope Benedict XVI, when he was still Cardinal Ratzinger and speaking on behalf of the Congregation for the Doctrine of Faith to the Catholic Bishops: "To choose someone of the same sex for one's sexual activity is to annul the rich symbolism and meaning, not to mention the goal's of the creator's sexual design" (reported in *The Times* (London), October 31, 1986). In other words, God is behind the moral standard, but he is providing a package deal – morality, behavior, physical nature. Note that I am not saying that I accept this particular line of argument, or that the Christian must accept it, or even that the Thomist must accept it. But to reject it, you must go at the notion of "natural" in this context. Either you must argue that, for humans, the natural far transcends the purely physical – or else reading a good novel would surely be unnatural – or you must argue that our modern understanding of sexuality shows that for humans the actual ways in which sexual acts are performed are much less important than the emotional bonds, or perhaps that modern biological theory and observation across species shows that homosexual acts are far less unnatural than we once thought. You cannot just simply say: "If it feels good, then it's ok." That may be your attitude, but it is not a moral attitude given natural law theory.

If you are a nonbeliever, then you don't need any of this. If you are a Christian, accepting the God of the last chapter, can you argue this way without violating science? I don't see why not, although do note – as the last paragraph implies strongly – you cannot proceed in ignorance or violation of science. There are those who would argue that the official Catholic prohibition on artificial methods of birth control is a paradigmatic instance of precisely this happening. If one is simply to think of human sexual intercourse as a matter of having babies – that the act of having sex must "retain its intrinsic relationship to the procreation of human life" – then obviously it would be wrong, under natural law theory, to use contraceptive devices. "[A]n act of mutual love which impairs the capacity to transmit life which God the Creator, through specific laws, has built into it, frustrates His design which constitutes the norm of marriage, and contradicts the will of the Author of life. Hence to use this divine gift while depriving it, even if only partially, of its meaning and purpose, is equally repugnant to the nature of man and of woman, and is consequently in opposition to the plan of God and His holy will" (Paul VI 1968, 13). (For obvious reasons, the rhythm method might be judged natural and hence be permissible.) However, if you start to see that sexual intercourse in the mammals, especially in the animals closest to us (the bonobos), plays a broader role, promoting social links, then the situation changes very quickly. Intercourse and reproduction are not quite the fused package deal that many think they are. Put into the pie the fact that today, given modern medicine and similar boons of civilization, far fewer children die than previously, with the consequent potential for very large families, a strain on individuals and on society, and it is far from obvious that natural law theory prohibits artificial birth control. Sex for sex's sake is an important factor in promoting family bonds, and it should not come at the expense of adding stress through ever-larger numbers of children. At least, if you allow that modern medicine and improved nutrition and sanitation are permissible, then it is far from obvious that artificial birth control is not permissible.

I don't want to get caught up here with arguments about either homosexuality or birth control. I am using these examples to show the nature of natural law theory as well as how informed thinking is needed to make sure that there are no clashes with science. What I want to show is that natural law theory is a sophisticated theory that makes sense of God, humans, and the role of morality in our relationship to God. It is a theory that in itself does not force you into conflict with science. Of course, for the Christian, there is a bit more to the story than this. God is good, we are created good, and yet we sin. Even the most secular of us must recognize that, even though there is morality, we are not always moral. One can give naturalistic explanations. As a Darwinian, I would argue that we are a mélange of the altruistic and the selfish – we have to be altruistic in order to be social, but we have to look after ourselves, and our genes are always on the lookout for the main chance (Ruse 2001). Others – for instance Freudians – have different explanations, but we all must have explanations. Christians are obviously not unaware of our propensity to sin. Saint Paul, expectedly, is pretty hot on the topic. "For we know that the law is spiritual; but I am carnal, sold under sin. For that which I do I allow not; for what I would, that do I not; but what I hate, that I do" (Romans 15:14–15). The classic Christian explanation is that although God created us good, we have fallen into sin – Genesis gives the story of Adam, Eve, the serpent, and the apple. My suspicion is that, literalists apart, there is something of a division among Christians today on this issue. Some look to a particular act or acts; others are more inclined to note that this is all part of human nature as given to us by biology and culture. These latter would accept the secular arguments entirely but want to put them in a theological context. The point again is that one can hardly describe Christians as being naïve about human nature – indeed, they often complain that post-Enlightenment secularists are the ones who are naïve – and that even if you do not accept the Christian position, there is no good reason to sneer at those who do.

This brings us back, one last time, to substantive ethics. Here we have to be wary of a certain propensity to circularity. Even the most secular of us are working against a background of two thousand years of Christianity, so what we think is moral and right is obviously arrived at in an atmosphere of Christian thinking on the subject. At least this is true of just about everyone except a philosopher like Nietzsche, who thought he could break with the past, although, as Dostoevsky showed in *Crime and Punishment* (featuring his would-be Nietzschian, the murderer Raskolnikov), it is not always easy for even the strongest of us to break with the norm. My point is that we should not expect to find much difference between secular accounts of substantive ethics and Christian accounts, and by and large we do not. Does this mean that all of us can and should turn to the Bible for moral guidance? What about the trolley problem or the *Bleak House* effect? My suspicion is that these are issues to wrestle with for the Christian as much as for the secular naturalist. Certainly, a lot of Christians would think that personal relationships count heavily and that loving your neighbor means family first. And don't forget that there is flexibility and ambiguity in Christianity – more charitably, room for discussion. Jesus tells us to leave our families. ("If any man come to me, and hate not his father, and mother, and wife, and children, and brethren, and sisters, yea, and his own life also, he cannot be my disciple." Luke 14:26) Paul tells us to care for our families, to love our spouses, to cherish our children, and to respect our parents. ("Children, obey your parents in the Lord: for this is right. Honour thy father and mother; which is the first commandment with promise; That it may be well with thee, and thou mayest live long on the earth." Ephesians 6:1–3) My sense is that most Christians would say that their religion helps them to deal with knotty moral problems, without necessarily offering simple rules. There is nothing there that tells us we must fly in the face of science.

SOULS

Christians claim that humans are made in the image of God and that this means that we have souls. Let us begin by digging into the meaning of this claim and then see how our discussion relates to science. Plantinga again is helpful in reminding us here that although the Christian is going to be very interested in what science (and its secular philosophers) say about the mind – "One of the principal aspects or elements of this image [of God] is precisely this ability to think" (1999, 19) – it is the soul and its connection to God that is the topic at issue. The scientist might think this all irrelevant, but in a way that is exactly the point. The starting line for Christians, even for the most nonliteralist of Christians, is Genesis. Animals and plants have had the breath of life breathed into them, and humans are uniquely different because they are made in the image of God. Genesis 1 and Genesis 2 are definitive on this: "And God said, Let us make man in our image, after our likeness: and let them have dominion over the fish of the sea, and over the fowl of the air, and over the cattle, and over all the earth, and over every creeping thing that creepeth upon the earth. So God created man in his own image, in the image of God created he him; male and female created he them" (Genesis 1:26–7). "And the Lord God formed man of the dust of the ground, and breathed into his nostrils the breath of life; and man became a living soul" (Genesis 2:7).

What is meant by this? Start with the soul part. Jewish thinking, certainly early Jewish thinking, was never into mind-body dualism. It was always more a matter of a vital principle that animates clay and makes it living. The great twentieth-century American theologian Reinhold Niebuhr admits that the "Biblical doctrine that man was made in the image of God and after His likeness is naturally given no precise psychological elaboration in the Bible itself." We are not to look here for sophisticated philosophy. "Nor does Biblical psychology ever achieve the careful distinctions of Greek thought. As in early Greek thought, spirit and soul are not at first carefully distinguished in the Bible. *Ruach* and *nephesh*, both meaning 'breath' and 'wind,' are used interchangeably in the Old Testament and they connote the Hebraic sense of the unity of body and soul rather than any special idea of the transcendence of spirit" (Niebuhr 1941, 151). Although there were greater Greek influences (more on these in a moment) in the New Testament, still it is the unity of body and mind that counts for the understanding of soul. Paul, for instance, never thinks of the resurrection of individual humans except in the context of bodily resurrection. "So also is the resurrection of the dead. It is sown in corruption; it is raised in incorruption: It is sown in dishonour; it is raised in glory: it is sown in weakness; it is raised in power. It is sown in a natural body; it is raised in a spiritual body. There is a natural body, and there is a spiritual body" (1 Corinthians 15:42–44).

What about being made in the image of God? Everyone seems to agree that, at the least, this implies intellectual abilities. Saint Augustine was definitive on this.

It is in the soul of man, that is, in his rational or intellectual soul, that we must find that image of the Creator which is immortally implanted in its immortality.... Although reason or intellect be at one time dormant within it, at another appears to be small and at another great, yet the human soul is never anything but rational and intellectual. Hence if it is made after the image of God in respect to this, that it is able to use reason for the understanding and beholding of God, then from the very moment when that nature so marvellous and great began to be, whether this image be so worn down as to be almost none at all, whether it be obscure and defaced or bright and beautiful, assuredly it always is. (Niebuhr 1941, 154–5, quoting Augustine, *De Trinitate*, XIV, 4, 6)

Karl Barth concurs. At some level, we have to have the ability to know God, to be aware of ourselves as beings, and to understand our relationship to God. Brute animal force cannot do this. He writes: "Man has Spirit, and through the Spirit is the soul of his body. This means at least that, by reason of his creaturely being, he is capable of meeting God, of being a person for and in relation to Him, and of being one as God is one." What does this mean? Man "is capable of being aware of himself as different both from God and from the rest of the created world. He is capable of recognizing himself of being responsible for himself. He exists in the execution of this self–recognition and self-responsibility before his creator" (Barth 1960, 395).

Being made in the image of God also implies a moral sense: the understanding of the difference between right and wrong and the understanding that we should do right and never do wrong. This is why human free will, discussed in the previous chapter, although problematic, is so very central to the Christian picture. It is all a matter of responsibility, responsibility ultimately to God, as is pointed out by Barth's fellow Swiss theologian Emil Brunner. "The fact that man must respond, that he is responsible, is fixed; no amount of human freedom, nor the sinful misuse of freedom, can alter this fact. Man is, and remains, responsible, whatever his personal attitude to his Creator may be. He may deny his responsibility, and he may misuse his freedom, but he cannot get rid of his responsibility. Responsibility is part of the unchangeable structure of man's being" (Brunner 1952, 56–7). What God has created is good, so human nature is essentially good, but for the Christian we are fallen – hence the need of the Incarnation, the Crucifixion, and the Resurrection – and this taints our nature. The important point here is that mind is certainly included in soul, but it is only one aspect. At a minimum, it has to be mind infused with freedom and morality and responsibility – an entity capable of and directed toward love, for all that we are limited and corrupted.

Now we have to start setting things in historical context. When Jesus died on the Cross, when he rose and left for heaven, there was no Christian religion. This was the work – articulating doctrines and setting up social structures – started by the disciples and the early apostles, notably Paul. It was very much the work of the early church authorities and theologians, the "church fathers," especially Augustine (354–430). It has been continued to the present, with Catholics receiving the greatest boost in the writings of Saint Thomas Aquinas. Protestants got their boost from the great reformers, especially Martin Luther and John Calvin. It is in this context that the thinking of the great Greek philosophers, particularly Plato (and the neo-Platonists) and Aristotle, becomes pertinent. It is this that was drawn on extensively by Christian thinkers.

Not that all was uniform. We have long had hints that the Greek approaches to soul are different, and the differences resonate even today. Plato's best-known treatment of the topic comes in his *Republic*, where, famously, he argues that the soul has three parts: the intellectual part; the courageous or spirited part; and the appetitive part, the part that fuels the body. Plato is no ascetic – he does not want to deny the importance of the lower parts – but he is clear that the well-functioning human is one ruled by the intellect. Both in the *Republic* and in the *Phaedo*, the dialogue dealing with the last day of Socrates, it is made clear that in some sense the soul – the intellectual part, certainly – is separate from the body, and, most importantly, does not die when the body dies. In other words, Plato (like Descartes to follow him) is a dualist. The human soul is immortal. (Plato believes in preexistence as well as postexistence). Having governed the body during one's lifetime, the soul now goes on to a better place.

Aristotle's thinking on the subject, particularly in his *De Anima* – thinking known technically as "hylomorphism" – is at one with his thinking about causation. It will be remembered that there are four kinds of cause: efficient cause, what gets things done; material cause, the substance; formal cause, the plan or idea; and final cause, the intended result or goal. For Aristotle, soul is the formal cause of the body. In other words, soul is to the body as the idea of a sculpture of Hermes (say) is to that bronze from which it is made.

> Among substances are by general consent reckoned bodies and especially natural bodies; for they are the principles of all other bodies. Of natural bodies some have life in them, others not; by life we mean self–nutrition and growth and decay. It follows that every natural body which has life in it is a substance in the sense of a composite.
>
> Now given that there are bodies of such and such a kind, viz. having life, the soul cannot be a body; for the body is the subject or matter, not what is attributed to it. Hence the soul must be a substance in the sense of the form of a natural body having life potentially within it. But substance is actuality, and thus soul is the actuality of a body.... (*De Anima* 412b12–22, Barnes 1984, 656)

Be careful, however, of your reading of the word "substance." Soul for Aristotle is not a substance as it is for Plato (or for Descartes two millennia later). It is the forming of substance in some wise and hence cannot just be split off from the bodily. "From this it is clear that the soul is inseparable from the body, or at any rate that certain parts of it are (if it has parts) – for the actuality of some of them is the actuality of the parts themselves" (657:413a4–6).

Aristotle is a biologist, so he is sensitive to the living world around him, and he is also a Greek, so there is no nonsense about where we humans stand in the scale of things. There are those organisms with just the basic kind of soul, plants particularly; then there are those a grade above, animals; and finally there are those with the rational capacity, humans. "Of the psychic powers . . . some kinds of living things . . . possess all, some less than all, others one only. Those we have mentioned are the nutritive, the appetitive, the sensory, the locomotive, and the power of thinking. Plants have none but the first, while another order of things has this plus the sensory." And some are above this. "Certain kinds of animals possess in addition the power of locomotion, and still others, i.e. man and possibly another order like man or superior to him, the power of thinking and thought" (659–60:414a29–414b19).

Platonic-type thinking had great influence on Augustine, but it was Aristotle's influence on Saint Thomas Aquinas that had the more lasting impact. Indeed, to this day Catholic thinking about the soul is much indebted to Aristotle. Soul and body cannot be ripped apart, for the one animates and informs the other. Since 1869, when Pope Pius IX decreed that human life begins at conception – a decision not entirely unrelated to the need to harmonize with his previously (1854) declared dogma that the Virgin Mary was "immaculately conceived" and hence free from sin from the zygote on – this has led to the embarrassing fact that Aquinas is out of step with the church's thinking (Pasnau 2001; but see Haldane and Lee 2003). In Aristotle's footsteps, Aquinas thought that the thinking soul develops only gradually in the human from a lesser kind of vegetative and then animal soul, and hence he did not think that abortion in the early stages of pregnancy is murder. (Saints Jerome and Augustine would have agreed on this. Of course, all of them thought that abortion is morally wrong.)

Protestant thinkers tend to emphasize the more biblical aspects of thinking about soul and want to distinguish themselves from Catholic theology and its influences. Barth, for instance, is rather rude about Plato and Aristotle. (A major reason being that he wants nothing of secular proofs of the immortality of the soul. Anything of this nature must be the direct gift of God.) But it is hard to imagine that Barth could have thought as he did without these great pagan thinkers, Aristotle particularly. He divides up living things in a suspiciously familiar fashion. All physical bodies are spatial, but not all spatial entities are alive. And not all that are alive are as high up the scale as others. "A plant would never live in an independent manner; to do so it would have to be transformed into a very different material body." Plants are above stones, but there are others above them. "[T]he animal body, and distinctly and recognizably the human, will always have at least the possibility of independent life, of being the body of a soul. When a material body is besouled, it does not cease to be a material body, but only to be merely material body" (Barth 1960, 377). More than this, we see the integration of soul and body, especially in Jesus Christ, who is for Barth the epitome, the apotheosis, of what it is to be truly human. "He is one whole man, embodied soul and besouled body: the one in the other and never merely beside it; the one never without the other but only with it, and in it present, active and significant; the one with all its attributes always to be taken as seriously as the other. As this one whole man, and therefore as true man, the Jesus of the New Testament is born and lives and suffers and dies and is raised again" (Barth 1960, 327).

Soul is not spirit. The latter is very much akin to what Quakers call the inner light or that of God in every person. It is the force from God that animates and maintains us. But soul equally is not something to be distinguished physically from body. The two are as one.

> The soul is not a being for itself, and it cannot exist for itself. Soul can awake and be only as soul of a body. Soul presupposes a body whose soul it is, i.e., a material body which, belonging to soul, becomes an organic body. Soul is inner – how could it be this if it had no outer? Soul is movement in time – how could it be this if it did not have an inalienable spatial complement, if it had no place? Soul fulfills itself in specific perceptions, experiences, excitations, thoughts, feelings and resolutions – how could it do this if it had no means in and through which it could exhibit itself? But all of these, outwardness and space and means, it does not have of itself. All these constitute its body. Thus in being soul, it is not without body. It is, only as it is soul of a body. (Barth 1960, 373)

In short, traditionally, the Christian soul is part of the very life force of humans, something made special by our intellect and moral capacity.

AND SCIENCE

What are we to say about all of this in the light of our inquiry? The question is not whether you accept this as true. I presume that if you are not a Christian you reject a lot of it. But this is not the point of our inquiry. The question is whether, on the basis of modern science, you should reject it. This is a much different question, and there are many who would say that you should. Obviously, this whole discussion is predicated on the belief that we humans are special. We may not be the only special beings – there are debates about this – but we are a central part of the story. Presumably we could have had green skin and twelve fingers. We might not have been bipedal, although Saint Thomas thought this was better than the alternatives. I am not sure about sex. But humanlike beings – beings with our intellectual and moral capacities – had to exist. Yet, as we saw in the Introduction, Stephen Jay Gould argued that this goes against our understanding of the evolutionary process: far from being necessary, consciousness is "a lucky afterthought" (Mehren 1989, E1). His sparkling work *Wonderful Life* (1989) hammered home this theme.

Some have tried to counter this. Although no friend of religion, Richard Dawkins (1986) thinks that intelligent beings would almost certainly have emerged as the result of biological arms races. Adaptations improve as the result of competition with other organisms, and eventually the best of all possible adaptations – intelligences – emerge. The paleontologist Simon Conway Morris (2003) argues that there is a hierarchy of ecological niches, ending with intelligence, and that sooner or later evolution would have produced humanlike beings: "if we humans had not evolved then something more-or-less identical would have emerged sooner or later" (p. 196). The physicist Robert John Russell (2008) puts direction in at the quantum level, arguing that God guides evolution and that He intended the appearance of *Homo sapiens*. My own inclination is to reject all of these suggestions. Arms races are controversial, and there is no strong reason to suppose that – on the one-shot opportunity here on earth – intelligence will necessarily emerge. Sophisticated brains are expensive in the sense of demanding high-quality fuel (like meat) to build and maintain. Sometimes it is better to be a bit dumb but easy to keep. There is little reason to believe that niches exist independently of organisms. Organisms themselves obviously have a major role in niche creation. Why is it inevitable that – again, given a one-shot opportunity – one kind of niche, the human kind, will emerge? Putting direction in at the quantum level sounds too much like a fancy form of Intelligent Design theory. Your imagination is not big enough to see what might have happened, so you invoke God. And as with Intelligent Design theory, there are horrendous theological problems. If God works to provide desirable variations, why did he not prevent undesirable variations, like those that lead to major genetic diseases? Once you bring God directly into the mixture, it is hard to relieve Him of direct responsibility.

Theological problems need theological solutions. My own favored suggestion is to follow Saint Augustine and stress that God lies outside time. "But, beloved, be not ignorant of this one thing, that one day is with the Lord as a thousand years, and a thousand years as one day" (2 Peter 3:8). As a necessary being, He is also outside space as we know it. What is pertinent is that evolution has produced humans; hence evolution could produce humans. That means, given enough trials, somewhere down the road humans would appear. What we don't know is how easy it is to start life and whether, once life starts, humans are nigh inevitable, likely, rare, or almost (but not totally) impossible. But this is irrelevant to God. Presumably, having created one universe, He could create an infinite number of universes, side by side or end to end. Unlike us, He is not waiting patiently for something to happen. It will! And that is all that is needed, no matter how random or undirected evolution may be. Given the laws that He created, humans were bound to appear.

This is not the end of the worries. Another part of the problem is that, for all that we have a fair amount of information now about the Christian notion of the soul, there are still fuzzy aspects – especially about the relationship between life itself and then the human thinking abilities. As we have just seen, traditionally soul has been understood (in part) as involving the life force. As it happens, given that for Aristotle the life force is less a thing, ethereal spirits driving things, and more something to be understood in terms of organization, in some respects his thinking is surprisingly pertinent. Today's biologists thinking about life would stress organization – the genetic code, the ways in which the many genes interact to make the organism – rather than some actual thing, some substance. The point is that today's biologists think that they are now getting on top of all of this – evolutionary development (evo devo) is showing a huge amount about how the organism is formed and worked – and so even though there are many gaps in our knowledge, it is reasonable to think (as I myself have argued earlier in this book) that life itself is natural, something within the mechanistic model of understanding. If this is so, then the following kind of sentiment is ruled out. "In the scientific view of Evolution, however, there are points which suggest that behind the mechanical, causal series, there lies a mystery; namely, the points where we cannot directly explain the 'consequence' from its 'antecedent'. Thus from the point of view of pure causality it is unintelligible – and in the highest degree improbable – that the highly complicated organic material with which life is connected, can have been formed out of accidental combinations of non-organic molecules of matter" (Brunner 1952, 40). This was written by Emil Brunner in the middle of the last century (referring to the ideas of the French Scientist Lecomte du Noüy), and it is obviously still the message of the Intelligent Design theorists (Behe 1996). However, apart from the fact that it is preferable not to think of natural selection as "accidental" – although it is certainly not teleological in any hard-line way – one doubts that the basic sentiment is still tenable. Life is no longer unintelligible in the way supposed in this passage.

Thus far, you might say that science pushes out (or is pushing out) the theological notion of soul, inasmuch as it refers to life as such. But as we have seen, the Christian notion of the soul for humans has always been more than just vitality. It brings in mind – thinking and moral action. If you are considering an Augustinian/Cartesian notion of soul, you are dealing with consciousness on its own. If you are dealing with an Aristotelian/Thomistic notion of soul, you are dealing with consciousness and life as it supports and gives rise to consciousness. Either way, considering this distinctive part of human life, science runs out of steam. Chalmers (2006) is right: if ever there was a place for talk of emergence, it is here – although remember that the very point of emergentists is to stress, even more than mechanists, that we reach the limits of our understanding. To repeat: "The existence of emergent qualities thus described is something to be noted, as some would say, under the compulsion of brute empirical fact, or, as I should prefer to say in less harsh terms, to be accepted with the "natural piety" of the investigator. It admits no explanation" (Alexander 1920, 2, 46–7). At this point, suppose the Christian says: "We cannot explain mind scientifically, but we know what it is. It is sentience. It is self-awareness. This is as much a reality as anything could be. And, as such, I want to relate it to God. God is sentient. God is self-aware. By and large, the rest of creation doesn't really measure up that well in this respect. This is what makes me created in His image." I don't see this as something particularly helpful scientifically. But that is not the point. Leaving the Churchlands and company to one side (on the grounds that their position is really not that plausible), the real point is that the scientist cannot stop the Christian from saying this. The scientist cannot say: "Oh, no. The mind is just molecules buzzing around. No more. Nothing else to be said." So what I am saying is that if the Christian wants to promote an understanding of soul in this sense – at least including mind, but probably also life as it makes for mind – then I don't see that the scientist can object. Not *qua* scientist, that is.

This is not to say that the Christian as a Christian is now out of the woods with respect to souls. There are still all sorts of problems about how you deal with babies, and even more with fetuses, with nonthinking humans. But I suspect that this is a concern for the Christian and not really for those of us thinking about the science-religion relationship. At least it should be, although it cannot be denied that there are many today who do want to muddy the waters. When you start asking about the origin of the human soul, Barth probably had the best idea when he said that he really could not care less where it came from. It was where it was going that was his big worry! However, there are those who want to get into the origin question. John Paul II (1997) was adamant that the origin of the soul had to be miraculous. Reluctantly, I suspect that if you really must insist on each human soul having been created miraculously (Creationism, in this sense), then since there is the gap in science with respect to sentience-intelligence and its relationship to the physical world (a gap that science cannot ever fill, not simply is not yet able to fill), then you could argue this. But it does seem an awfully large amount of work for God, and you might rather simply argue that the miracle lies in the fact that it happens, not in how it happens – a kind of natural supernaturalism advocated by Thomas Carlyle in *Sartor Resartus*. Some form of Traducianism (that souls are a legacy of the last generation) seems to fit better with an evolutionary perspective. Expectedly, given the unfurling implications of his theology, there is some precedent in Augustine for a Traducian solution. (He was divided on the issue, but the merit of the Traducian option is that it seems to explain at once why souls are tainted with sin, whereas the Creationist option has to explain how something newly created by God is nevertheless tainted. See Mendelson 1995.)

What does seem true is that those who insist that souls come with zygotes, way before sentience and serious thinking, are muddying the waters even further. At the least you have to start raising claims about potentiality, and these are notoriously tricky. Consider the recent discovery that the human zygote can divide three times and, if separated, all of the eight new cells can make a fully functioning human being (Maienschein 2003). How many souls, therefore, do we have with any single first cell? One, or potentially eight, or actually eight? Somehow eight, potential or actual, seems a bit irrelevant. Why not say rather that we have potentially one until the sentience and thinking kicks in? This problem is not going to be solved here and now. And in the future, appeals to science are not going to solve these sorts of issues. They are theological claims and as such must be solved theologically. One suspects that Aristotle may have been rejected a little too quickly.

ESCHATOLOGY

There is life after death. This is the central message – at least, the central promise – of Christianity (Hick 1976). The ancient Jews did not have such a belief. When Abraham made his covenant with God, the promise was that his line would persist in large numbers. There was nothing about personal immortality. At most, the Jews had thoughts of something like the early speculations of the Greeks, a kind of gloomy Hades (the Jews called it Sheol) where souls persist in darkness, out of communion with God.

> O lord God of my salvation, I have cried day and night before thee: Let my prayer come before thee: incline thine ear unto my cry; For my soul is full of troubles: and my life draweth nigh unto the grave. I am counted with them that go down into the pit: I am as a man that hath no strength: Free among the dead, like the slain that lie in the grave, whom thou rememberest no more: and they are cut off from thy hand. (Psalms 88:1–5)

Earlier civilizations did have beliefs in a desirable afterlife – something of this nature may have been in circulation in Egypt in the third millennium BCE – but it was around four or five centuries BCE, in Greece and in Israel, that a more sophisticated and worthwhile eternity started to open its doors. Socrates in the *Phaedo* (and hence, supposedly, on the day of his death in 399 BCE) argues for a desirable existence once this mortal life is finished. It was in the same period or a bit later, perhaps influenced by Zoroastrian ideas, that such thinking started to spread among the Jews. There was a major difference however, one that stems from differences seen and discussed in the previous sections. Whereas Plato was arguing for a soul without body that exists after this life here on earth – a necessary consequence of his belief that this world of ours is the world of change and corruption, whereas the soul is going to last eternally (in the world of the Forms and of mathematics) – Jewish thought always saw an integration of mind and body (in other words, it was more in tune with Aristotle's thinking, although surely independently arrived at). The Book of Daniel (set in the Babylonian captivity, beginning 586 BCE) gives the first hint of a personal resurrection. "And many of

them that sleep in the dust of the earth shall awake, some to everlasting life, and some to shame and everlasting contempt" (Daniel 5:2).
Note that already we are getting what is going to be an important
part of the Christian message (something that is found back with the
early Egyptians), namely, that there will be a judgment of how lives
were lived here on earth, and only those who merit it will get the
good eternal life. "And they that be wise shall shine as the brightness
of the firmament; and they that turn many to righteousness as the
stars for ever and ever" (Daniel 5:3).

By the time of Paul (around 50 CE) and the writing of the Gospels
(70 CE and fifty years after), Greek thought was well known in Israel,
and one does sometimes get language about eternal life that sounds
almost Platonic. For instance, when Jesus tells the parable of the rich
man: "But God said unto him, Thou fool, this night thy soul shall
be required of thee; then whose shall those things be, which thou
hast provided?" (Luke 12:20). (Consider also Isaiah 26:9. "With
my soul have I desired thee in the night; yea, with my spirit within
me will I seek thee early; for when thy judgments are in the earth,
the inhabitants of the world will learn righteousness.") However,
generally and persistently, the story of the New Testament is one of
the resurrection of the body – physical and spiritual are united in one
soul. This, of course, read literally, is the story of Jesus himself. On
the third day he rose – he rose physically and showed himself to his
followers, including Thomas, who doubted his triumph over death
and who was shown the marks of the Crucifixion. Before he left
to ascend to heaven, Jesus showed himself capable of walking and
talking and – for all that he seems to have been ethereal enough to
appear and disappear at will – even of eating to assuage what sounds
remarkably like hunger. ("And while they yet believed not for joy,
and wondered, he said unto them, Have ye here any meat? And they
gave him a piece of broiled fish, and of an honeycomb. And he took
it, and did eat before them" Luke, 24:41–43.) It is also the story
as promised to us. Saint Paul has the most sophisticated discussion
of this – expectedly, because he was the man with real theological
training. There will be a resurrection of the body; however, it will be
a spiritual body and not the physical one we possess today (or at the
point of death). Remember: "It is sown a natural body; it is raised a
spiritual body. There is a natural body, and there is a spiritual body"
(1 Corinthians 15:44).

There is the matter of qualifying for a good afterlife, not to mention the question of what exact form such a good afterlife takes. There will be a Day of Judgment, when God will separate the sheep from the goats, those who are destined for paradise, whatever that might be, and those destined for hell, whatever that might be. The final book of the Bible, Revelation, has detailed information about the Day of Judgment and the events around it – so detailed, in fact, that Christians have split bitterly over the exact interpretation. Even more significant than these divides are those between Christians who think that faith alone is the ticket to salvation and Christians who think that works must enter the equation. We encounter here in full the Protestant-Catholic divide. On the one hand, there are those Christians who look back to the great Protestant reformers Luther and Calvin (and beyond them to Augustine), and who argue that the key to salvation is faith and faith alone. Their text is from Saint Paul, who was himself a man who was saved by faith and not because of anything he had done. "By what law? of works? Nay; but by the law of faith. Therefore we conclude that a man is justified by faith without the deeds of the law" (Romans 3:27–28). This is not to say that good works are unimportant. No one labored for his Lord more than Paul. It is to say that good works are a mark that one has been saved, and in no way can they be a payment for salvation. Catholics and some others – Quakers, for instance, are part of the radical reformation and have always turned to the Holy Spirit for guidance rather than to the word of scripture – go the other way, arguing that good works do count and that it is by these that we shall be judged. Apart from the many exhortations of Jesus to do good to others, their text, one that Luther (expectedly) loathed, is found in the Epistle of James. "What doth it profit, my brethren, though a man say he hath faith, and have not works? can faith save him?" (2:14) Continuing: "Even so faith, if it hath not works, is dead, being alone" (2:17). Expectedly, by those that subscribe to justification by faith, this is known as a heresy, the Pelagian Heresy.

There are many more details and issues that could be raised here. We might discuss the nature of hell, or the possibility of God extending his mercy and eternal life to all humans and not just to a select few, or the prospects of salvation for children and others (including those born before Jesus or those born in lands unaware of the Christian message). But, whatever the relative importance of these

and like issues, for our purposes enough of the basic Christian picture of future things and hopes has now been presented. Remember,
the question is not whether we ought to accept it. Rather, the question is whether someone who accepts modern science has any good
reason to reject it purely on the grounds of science. Of course, the
scientist might well want to deny the possibility of life after death,
whether as a disembodied mind or as a resurrected body. The question is whether one can do this on the grounds of science. If the claim
were being made that, say, somewhere elsewhere in the universe we
shall find Saint Paul and Julius Caesar and Napoleon and Charles
Darwin – as minds alone or with bodies also – then, as a scientist,
one might be skeptical. But this is not the claim. It is rather that there
is another dimension of existence where resurrected bodies exist –
or minds, if that is all. It is the place of the spiritual body. As such,
I doubt that science can lay a finger on the idea. The scientist may
not much care for the religious claims – the philosopher Bernard
Williams (1973) once wrote an article on "the tedium of immortality" – but he or she is not able to reject them on the grounds that
they go against scientific understanding.

MYSTERY

The more one thinks about Christianity, the more the problems seem
to multiply. How on earth can a necessary being be a thinking being?
How can God be outside time and yet have emotions of love and
concern? How can God have left so many people outside the culture
in which he is known and cherished? How can God, a father, truly
value Hitler's free will over the suffering of Anne Frank at Bergen-
Belsen? The Christian may have no explanations, but the Christian
has an answer. God is infinite. We are finite. We get at most a half
picture, images through shards rather than the full views. "For now
we see through a glass, darkly; but then face to face: now I know in
part; but then shall I know even as also I am known" (1 Corinthians
13:12). Remember Pope John Paul II: "It should nonetheless be kept
in mind that Revelation remains charged with mystery." Hence,
"our vision of the face of God is always fragmentary and impaired
by the limits of our understanding" (John Paul II 1998, section 13).
There is nothing in science that suggests otherwise, and indeed (as we
have seen) much in an evolutionary perspective on human nature to
suggest precisely that our understanding is, and on this earth always

will be, less than complete.` The scientific position is not confirming the Christian claim, but it is certainly making it possible. There are those of us who think that the discoveries and theories in physics of the past hundred years – not to mention the possibility that problems like the body-mind relationship may be beyond our scope – should make us all a little more humble when it comes to existence, and that, even if we disagree, a little reserve might not be inappropriate.

Or is this truly so? Daniel Dennett is scathing. He thinks that Christianity happily embraces contradictions and, in so doing, takes one out of reasonable discourse – certainly takes one out of reason as science knows it – and hence is worthy of condemnation by science, even though science itself may not be speaking to the issues (obviously for very good reasons). Hence, here, the reach of science, or rather the reach of the standards of science, makes the Christian religion an exercise in futility. Dennett characterizes philosophical theology using a metaphor of the Canadian philosopher Ronald de Sousa. It is "intellectual tennis without a net." Dennett invites us to play with him.

> It's your serve. Whatever you serve, suppose I return service rudely as follows: "What you say implies that God is a ham sandwich wrapped in tinfoil. That's not much of a God to worship!" If you then volley back, demanding to know how I can logically justify my claim that your serve has such a preposterous implication, I will reply: "Oh do you want the net up for my returns, but not for your serves? Either the net stays up or it stays down. If the net is down, there are no rules and anybody can say anything, a mug's game if ever there was one. I have been giving you the benefit of the assumption that you would not waste your own time or mine by playing with the net down." (Dennett 1995, 154)

Dennett then explains his thinking in a somewhat less metaphorical fashion.

> Now if you want to *reason* about faith, and offer a reasoned (and reason–responsive) defense of faith as an extra category of belief worthy of special consideration, I'm eager to play. I certainly grant the existence of the phenomenon of faith: what I want to see is a reasoned ground for taking faith seriously as a *way of getting to the truth*, and not, say, just a way people comfort themselves and each other (a worthy function that I do take seriously). But you must not expect me to go along with your defense of faith as a

> path to truth if at any point you appeal to the very dispensation you are supposedly trying to justify. (Ibid.)

He continues by asking if we truly want to give up reason if reason has backed us into a corner. Is this really now the time for faith? He asks us if we would really be "willing to be operated on by a surgeon who tells you that whenever a little voice in him tells him to disregard his medical training, he listens to the little voice?" Dennett graciously realizes that in polite society we tend not to challenge people's deeply held beliefs, however nutty or dangerous they may be. "But we're seriously trying to get at the truth here, and if you think that this common but unspoken understanding about faith is anything better than socially useful obfuscation to avoid mutual embarrassment and loss of face, you have either seen much more deeply into the issue than any philosopher ever has (for none has ever come up with a good defense of this) or you are kidding yourself" (pp. 154–5).

Everything Dennett says is right. Everything Dennett says is irrelevant. The whole point is that the Christian cherishes reason as a gift from God. "Faith and reason are like two wings on which the human spirit rises to the contemplation of truth; and God has placed in the human heart a desire to know the truth – in a word, to know himself – so that, by knowing and loving God, men and women may also come to the fullness of truth about themselves (cf. *Ex* 33:18; *Ps* 27:8–9; 63:2–3; *Jn* 14:8; *1 Jn* 3:2)" (John Paul II 1998, Prologue). It is not a question of faith stepping in when we do not much like the findings of reason. Faith comes into play when reason and evidence are outstripped; but even then, reason is not abandoned. If we are to make sense of faith claims – the task of this and the previous chapter – then (remember John Locke) reason still rules. At least, it still has its essential role to play. You cannot, for instance, simply dismiss the problem of evil as a nasty trick of reason, irrelevant for the believer. You have still got to wrestle with it, as indeed Christians do. You have got to try to make some sense of evil within the story of creation by a good God and the subsequent actions and fate of humankind. That is why Christians turn to the Bible, to the Book of Job, for example. That is why Christians turn to their great thinkers, Augustine in *The City of God,* for example. It is true, as we have seen, that there has been a strand in Christian thought (universal possibilism) suggesting that God may even be able to change the laws of logic and mathematics. But we have seen also that Christians hardly embrace this as a license to argue or claim anything. More, there is a retreat to baffled silence. In the face of such a God, it is hard to know if one can say anything at all.

What of Dennett's objection that he wants an argument for the validity of faith, taking us beyond the nasty suspicion that it is simply a matter of self-deception about things we wish fervently were true? I suspect we have one of those circular situations here, akin to the circular situations Thomas Kuhn (1962) felt we have when it comes to justifying paradigms. It is hard to get out of your own circle in order to critique the other side. The nonbeliever is obviously going to argue that faith is self-deception, and may well back up this belief with a mass of psychological evidence. The believer is perhaps (almost certainly) going to agree that there may well be psychological evidence bearing on faith, but will claim nevertheless that faith is self-validating. If you have faith, then no proof is necessary. (This is an argument Plantinga often uses; although, as we have seen, he might argue that reason, not faith, is at issue.) I am not sure that this is all that can be said. The nonbeliever may still want to challenge the believer on theological grounds. But I suspect that this is all that can be said on scientific grounds.

There are, of course, those who would want to challenge people like Dennett in a more robust way. I refer to those enamored of natural theology, the belief that one can arrive at knowledge of God independently of faith, through reason and evidence. To them, the very starting point is that religion embraces the facts of experience and the rules of reason, perhaps appropriating science itself to its needs. Thus, for instance, there is a modern-day group of physics enthusiasts who think that the science shows that the constants of the universe are so precisely necessary for the support of life that none of this could have happened by chance (Barrow and Tipler 1986). I want to make it clear that my position does not in any way depend on the success (or the failure) of natural theological arguments like this. I am not going to offer a theoretical argument against natural theology, although as it happens I am not very keen on it, like many today (Ruse 2003). Scientifically, it seems to me to open itself too readily to refutation or belittling. For instance, the argument just offered (usually known as embracing the Anthropic Principle about the necessity of nature's units) seems to me to make unwarranted assumptions about what precisely the constants must be to sustain a functioning life-containing universe (Weinberg 1999). Theologically,

I sympathize with the great nineteenth-century Danish figure Søren Kierkegaard, who argued that faith is genuinely faith only if it requires a kind of commitment beyond the evidence, a commitment that natural theology by its very aim is trying to undermine.

And even if you do not share this second feeling – and in fairness I should point out that the Catholic Church had and still has a fondness for natural theology (while maintaining that faith must always be primary) – I am inclined to agree with John Henry Newman, the great Englishman who started life as an evangelical Anglican, moved to the High Church and leadership of the Oxford Movement, and finally went over to Rome, ending his long life as a cardinal. Of the then very popular Argument from Design, he wrote: "I believe in design because I believe in God; not in a God because I see design" (Newman 1973, 97). He continued: "Design teaches me power, skill and goodness – not sanctity, not mercy, not a future judgment, which three are of the essence of religion." Somehow, too often, natural theology points one in the wrong direction. I agree also with many Christians today who feel that it is too close to God-of-the-gaps theology. This occurs when one uses God to fill in when the going gets rough, rather as Newton used God to jiggle the universe a bit when the laws of nature got things out of focus and (I would say) as the so-called Intelligent Design theorists call on God to make complex organic parts when they think natural selection is not up to the job. This is bad science and bad theology.

None of this denies that once having made faith commitments, it is legitimate to turn to science to make sense of them. One can turn to science as one turns to anything else. Richard Swinburne draws our attention to the nature of light. He writes: "Now the situation in Quantum Theory is a two–model situation. The postulated entities, such as photons, the units of light, have some of their properties in common with waves and some with particles and there is no one model which we can provide of them. There is around us no familiar phenomenon which resembles the postulated photons more than either of these, and they resemble it very much equally." He continues:

> If we have two models for a postulated entity, we have necessarily to use words in analogical senses if we are to describe that entity. For since it cannot resemble both of its models very closely (since they are different models), we can only describe it in terms of

the two models by giving extended senses to the words used for describing the models. The fact that we can give such an analogical interpretation of theories which do not admit of an interpretation with words used in ordinary senses is of great interest. For our ability to do this is clearly independent of the truth of Quantum Theory. If it can be done for Quantum Theory when we have grounds for believing it to be true, it can be done for some other theory when we have grounds for believing it to be true. (Swinburne 1977, 69–70)

If someone wants to seize on this and use it to illustrate or illuminate a difficult theological point – say, of the relationship between God the Father and God the Son – this is a perfectly legitimate thing to do, at least from the viewpoint of science. One might want to reject such a move as theologically not very helpful; the two situations perhaps are not so very similar; but, as Swinburne notes, no one is challenging quantum theory. One is using it as a guide, as a heuristic.

CONCLUSION

My argument in these last two chapters has been that, although the central claims of Christianity are still constrained by reason, and although there is still certainly the need (as with natural law theory) to make sure that one's religion-based claims harmonize with modern science, these central core claims by their very nature go beyond the reach of science. I do not say that you must be a Christian, but I do say that in the light of modern science you can be a Christian. We have seen no sound arguments to the contrary.

CONCLUSION

What is the relationship between Christianity and today's science? It is obvious that there is no simple or single answer to this question. If one is thinking of the Christianity of our absent friends the fundamentalists, the Creationists, then there is simply massive conflict. You cannot believe in a six-thousand-year-old earth, a six-day creation, a worldwide flood, and at the same time accept modern physics, modern biology, modern geology. But fundamentalism is not the only form of Christianity and has little lien on the traditional form of the religion. If you follow the route marked out by Augustine and Aquinas, by Luther and Calvin, then the answer is very different. The basic, most important claims of the Christian religion lie beyond the scope of science. They do not and could not conflict with science, for they live in realms where science does not go. In this sense, we can think of Christianity and science as being independent, and we can see that those theologians who have insisted on the different realms were right in their view of the science-religion relationship.

This is not to say that there is no relationship at all between Christianity and modern science. Given that Christianity is, after all, a religion about the nature of this world and the place of human beings in it, such would be a very odd state of affairs indeed. At the least, a great deal of negotiation has been needed (and most probably still is needed) to work out the boundaries between science and religion. This is not something that can be decided a priori, before inquiry begins, but needs constant assessment, especially as science unfurls and develops. Also, even when boundaries are found, science and religion reach across to each other. Christianity cannot simply ignore the rules and norms of science, especially the standards of reasoned argument. Conversely, it is expected and appropriate for Christianity to make claims about the world of experience, for instance, in the moral sphere. It is simply that when this does occur, as with the application of natural law theory, great care must be taken to see that the theological conclusions are infused with the findings of really up-to-date science. As the science changes, so also may these conclusions. A delicate balancing act is needed. Today, no one who takes modern science seriously is going to deny some form of organic evolution. It is surely legitimate, therefore, for a religious person to think about ways in which God might have created in such a fashion, and turning for insight to St. Augustine's thinking about how a non-temporal God might have created a time-bound universe is surely an open possibility. I engaged in precisely this kind of argument in the last chapter. On the other hand, identifying the creation with the Big Bang is fraught with problems, not the least of which is the possibility that there might have been something prior to our Big Bang.

In a like fashion, there still are, and probably always will be, some grey or contested areas about the domains of science and religion. Take miracles, for instance. Logically, you cannot deny the stance of those who embrace the order-of-grace option, arguing that miracles stand outside the order of nature and are performed by God as an end to our salvation. Water could turn into wine. However, quite apart from the theological issues – is God really a high-class vintner? – naturally, water simply does not turn into wine. To claim otherwise is to violate the norms of science. Hence, one might argue that if religion insists that this must be true, then it is encroaching illegitimately on the realm of science. Noting in passing that this is a conclusion shared not only by nonbelieving scientists but also by many Christians – who take seriously the points to be made in the next paragraph – here I will leave the matter unresolved. The answer clearly depends on the allowable scope of science. If it is insisted that the scope is the whole of the natural world, without exception, then the order-of-grace option is disallowed. If, however, it is agreed that theological demands can enter into this discussion – an all-powerful

Creator can do what he pleases – then the order-of-grace option has legs.

Is it a sign of weakness that it is almost always going to be Christianity that must accommodate itself to the findings of science? Once it was possible to read Genesis fairly literally, because that was the direction in which the science pointed. Now such a reading is illicit. Once many thought that Saint Paul's views on women, on homosexuals, on slavery were fully acceptable. Now, in the light of modern social science, all of these assumptions have been (and are still being) challenged and reevaluated. Things do not go the other way. No physicists working as physicists are going to be bothered by reinterpretations of the Trinity. There are good reasons why Christians do not and should not see this asymmetry as a sign of weakness. Apart from the fact that there will almost certainly always be the major areas into which science cannot move, remember that for Christians reason is one of God's greatest gifts, the sign that we are indeed made in His image. Science therefore is a sacred task. It is also a difficult and challenging task, befitting creatures of our nature. Only slowly and with much effort will we discover the true nature of our home. As we do, our understanding of God and His nature and His works will obviously likewise change and mature. John Henry Newman had the right idea on this. The essentials of the Christian faith were revealed two millennia ago. Since then, theologians and scientists have been working to show exactly how these essentials play out in the creation. Theological understanding is always on the move. It is not evolutionary in the Darwinian sense – you could never drop or modify the initial faith claims – but it is developmental in a very real way. (Historically, Newman, who was always interested in science, was much influenced by anatomist Richard Owen's thinking about archetypes, the underlying Platonic ground plans of organisms, and about how they become ever-more adaptive as they are incorporated in real organisms having to survive over time.)

One doubts very much that today's frenetic partisans, from science and from religion, are going to change their minds very much. But the challenge of seeing the proper relationship between science and religion is there, and, both politically and intellectually, it is an important challenge. The hope is that the ideas and conclusions of this book will inspire others to join with the author in working on the task before us.

BIBLIOGRAPHY

Alexander, S. 1920. *Space, Time and Deity (The Gifford Lectures at Glasgow, 1916–1918 in Two Volumes)*. London: Macmillan.

Angier, N. 2001. The Bush years: Confessions of a lonely atheist. *New York Times Magazine*, January 14, 34–8.

Aquinas, St. T. 1952. *Summa Theologica, I*. London: Burns, Oates and Washbourne.

1975. *Summa Contra Gentiles*. Translator V. J. Bourke. Notre Dame: University of Notre Dame Press.

Augustine 1873. *Lectures or Tractates on the Gospel According to St. John*. Edinburgh: T & T Clark.

[413–26] 1998. *The City of God against the Pagans*. Editor and translator R. W. Dyson. Cambridge: Cambridge University Press.

1998. *Confessions*. Translator H. Chadwick. Oxford: Oxford University Press.

Bailey, L. H. 1897. *The Survival of the Unlike: A Collection of Evolution Essays Suggested by the Study of Domestic Plants* (second edition). New York: Macmillan.

Barkow, J. H., L. Cosmides, and J. Tooby, editors. 1991. *The Adapted Mind: Evolutionary Psychology and the Generation of Culture*. New York: Oxford University Press.

Barnes, J., editor. 1984. *The Complete Works of Aristotle*. Princeton, N.J.: Princeton University Press.

Barrett, P. H., P. J. Gautrey, S. Herbert, D. Kohn, and S. Smith, editors. 1987. *Charles Darwin's Notebooks, 1836–1844*. Ithaca, N.Y.: Cornell University Press.

Barrow, J. D., and F. J. Tipler. 1986. *The Anthropic Cosmological Principle*. Oxford: Clarendon Press.

Barth, K. 1957. *Church Dogmatics. Volume II: The Doctrine of God. Part 1*. Editors G. W. Bromiley and T. F. Torrance. London: T & T Clark International.

[1949] 1959. *Dogmatics in Outline*. New York: Harper and Row.

1960. *Church Dogmatics. Volume III: The Doctrine of Creation, Part 2*. Editors G. W. Bromiley and T. F. Torrance. London: T & T Clark International.

Bayliss, W. M. 1915. *Principles of General Physiology*. London: Longmans, Green.

Behe, M. 1996. *Darwin's Black Box: The Biochemical Challenge to Evolution*. New York: Free Press.

Berkeley, G. 1710. *A Treatise Concerning the Principles of Human Knowledge*. Editor J. Dancy. Oxford: Oxford University Press.

Bernacerraf, P. 1973. Mathematical truth. *Journal of Philosophy* 70: 661–79.

Bilodeau, D. J. 1997. Physics, machines, and the hard problem. In *Explaining Consciousness: The 'Hard Problem'*, ed. J. Sheav, 217–36. Cambridge, Mass.: MIT Press.

Black, M. 1962. *Models and Metaphors*. Ithaca, N. Y.: Cornell University Press.

Blumenbach, J. F. 1781. *Über den Bildungstrieb und das Zeugungsgeschäfte*. Göttingen: Dietrich.

Bohm, D. 1980. *Wholeness and the Implicit Order*. London: Routledge.

Boyle, R. [1688]1966. *A Disquisition about the Final Causes of Natural Things*. *The Works of Robert Boyle*, editor T. Birch, 5: 392–444. Hildesheim: Georg Olms.

[1687] 1996. *A Free Enquiry into the Vulgarly Received Notion of Nature*. Editors E. B. Davis and M. Hunter. Cambridge: Cambridge University Press.

Broad, C. D. 1925. *The Mind and Its Place in Nature*. London: Kegan Paul.

Brunner, E. 1952. *The Christian Doctrine of Creation and Redemption. Dogmatics, II*. Translator O. Wyon. Philadelphia: Westminster Press.

Bultmann, R. 1958. *Jesus Christ and Mythology*. New York: Charles Scribner's Sons.

Burgess, J., and G. Rosen. 1997. *A Subject with No Object: Strategies for Nominalist Interpretation of Mathematics*. New York: Oxford University Press.

Calvin, J. 1536. *Institutes of the Christian Religion*. Grand Rapids, Mich.: Eerdmans.

Carey, B. 2007. An active, purposeful machine that comes out at night to play. *New York Times*, October 23, Science sec., p. 1.

Carlson, E., and E. J. Olsson. 2001. The presumption of nothingness. *Ratio* 14: 203–21.

Carnap, R. 1956. Empiricism, sematics, and ontology. In his *Meaning and Necessity*, 205–21. Chicago: University of Chicago Press.

Carroll, S. B., J. K. Grenier, and S. D. Weatherbee. 2001. *From DNA to Diversity: Molecular Genetics and the Evolution of Animal Design*. Oxford: Blackwell.

Carson, R. 1962. *Silent Spring*. New York: Houghton Mifflin.

Chalmers, D. J. 1996. *The Conscious Mind*. New York: Oxford University Press.

— 1997. Facing up to the problem of consciousness. In *Explaining Consciousness – The 'Hard Problem'*, ed. J. Shear, 9–32. Cambridge, Mass.: MIT Press.

— 2006. Strong and weak emergence. In *The Re-Emergence of Emergence: The Emergentist Hypothesis from Science to Religion*, ed. P. Clayton, and P. Davies, 244–55. Oxford: Oxford University Press.

Chomsky, N. 1957. *Syntactic Structures*. The Hague: Mouton.

Churchland, P. M. 1995. *The Engine of Reason, the Seat of the Soul*. Cambridge, Mass.: MIT Press.

Churchland, P. S. 1997. The hornswoggle problem. In *Explaining Consciousness – The 'Hard Problem'*, ed. J. Shear, 37–44. Cambridge, Mass.: MIT Press.

Clark, A. 2000. *Mindware: An Introduction to the Philosophy of Cognitive Science*. New York: Oxford University Press.

Clarke, R. 2003. *Libertarian Accounts of Free Will*. New York: Oxford University Press.

Clayton, P., and P. Davies, editors. 2006. *The Re-Emergence of Emergence: The Emergentist Hypothesis from Science to Religion*. Oxford: Oxford University Press.

Cobb, J. B., and D. R. Griffin. 1976. *Process Theology: An Introductory Exposition*. Philadelphia: Westminster Press.

Coleman, W. 1971. *Biology in the Nineteenth Century: Problems of Form, Function and Transformation*. New York: John Wiley.

Conway Morris, S. 2003. *Life's Solution: Inevitable Humans in a Lonely Universe*. Cambridge: Cambridge University Press.

Cooper, J. M., editor. 1997. *Plato: Complete Works*. Indianapolis: Hackett.

Crick, F. 1988. *What Mad Pursuit: A Personal View of Scientific Discovery*. New York: Basic Books.

— 1994. *The Astonishing Hypothesis: The Scientific Search for the Soul*. New York: Charles Scribner's Sons.

Curd, P., and R. D. McKirahan, editors. 1996. *A Presocratics Reader*. Indianapolis: Hackett.

Cuvier, G. 1817. *Le règne animal distribué d'aprés son organisation, pour servir de base à l'histoire naturelle des animaux et d'introduction à l'anatomie comparée*. Paris: Deterville.

Darwin, C. 1845. *Journal of Researches into the Natural History and Geology of the Countries Visited during the Voyage of H.M.S. Beagle round the World* (second edition). London: John Murray.

1859. *On the Origin of Species by Means of Natural Selection, or the Preservation of Favoured Races in the Struggle for Life*. London: John Murray.

1862. *On the Various Contrivences by which British and Foreign Orchids are Fertilized by Insects, and the Good Effects of Intercrossing*. London: John Murray.

1868. *The Variation of Animals and Plants under Domestication*. London: John Murray.

1871. *The Descent of Man, and Selection in Relation to Sex*. London: John Murray.

1872. *On the Origin of Species* (sixth edition). London: John Murray.

1879. Preliminary Notice. In E. Kraus, *Erasmus Darwin*, 1–127. London: John Murray.

1958. *The Autobiography of Charles Darwin, 1809–1882*. Editor Nora Barlow. London: Collins.

1985-. *The Correspondence of Charles Darwin*. Cambridge: Cambridge University Press.

Davies, P. 1999. *The Fifth Miracle: The Search for the Origin and Meaning of Life*. New York: Simon and Schuster.

2006. The physics of downward causation. In *The Re-Emergence of Emergence: The Emergentist Hypothesis from Science to Religion*, ed. P. Clayton and P. Davies, 35–52. Oxford: Oxford University Press.

Dawkins, R. 1976. *The Selfish Gene*. Oxford: Oxford University Press.

1983. *Universal Darwinism: Molecules to Men*, ed. D. S. Bendall. Cambridge: Cambridge University Press, 403–25.

1986. *The Blind Watchmaker*. New York: Norton.

1995. *A River Out of Eden*. New York: Basic Books.

1997. Is science a religion? *The Humanist* 57(1), 26–9.

2003. *A Devil's Chaplain: Reflections on Hope, Lies, Science and Love*. Boston and New York: Houghton Mifflin.

2007. *The God Delusion*. New York: Houghton Mifflin.

Deacon, T. W. 2006. Emergence: the hole at the wheel's hub. In *The Re-Emergence of Emergence: The Emergentist Hypothesis from Science to Religion*, ed. P. Clayton and P. Davies, 111–50. Oxford: Oxford University Press.

Dennett, D. C. 1984. *Elbow Room: The Varieties of Free Will Worth Wanting*. Cambridge, Mass.: MIT Press.

1995. *Darwin's Dangerous Idea*. New York: Simon and Schuster.

1997. Facing backwards on the problem of consciousness. In *Explaining Consciousness – The 'Hard Problem'*, ed. J. Shear, 33–6. Cambridge, Mass.: MIT Press.

Dennett, D. C., and R. Winston. 2008. Is religion a threat to rationality and science? *The Guardian*, April 22, 14.

Descartes, R. [1644]1955. *The Principles of Philosophy. The Philosophical Works of Descartes*, translators E. Haldane and G. R. T. Ross, 1: 201–302. New York: Dover.

1964. *Discourse on Method.* In *Philosophical Essays*, trans. L. Lafleur, 1–57. Indianapolis: Bobbs-Merrill.

1964. Meditations. In *Philosophical Essays*, trans. L. Lafleur, 59–143. Indianapolis: Bobbs-Merrill.

1964. *Rules for the Direction of the Mind.* In *Philosophical Essays*, trans. L. Lafleur 145–236. Indianapolis: Bobbs-Merrill.

[1664] 1972. *Treatise of Man.* Trans. T. S. Hall. Cambridge Mass.: Harvard University Press.

Diderot, D. 1943. *Diderot: Interpreter of Nature.* New York: International Publishers.

Dijksterhuis, E. J. 1961. *The Mechanization of the World Picture.* Oxford: Oxford University Press.

Dobzhansky, T. 1937. *Genetics and the Origin of Species.* New York: Columbia University Press.

Dronke, P., editor. 1988. *A History of Twelfth-Century Western Philosophy.* Cambridge: Cambridge University Press.

Edwards, P. 1967. Why. In *The Encyclopedia of Philosophy*, ed. P. Edwards, Vol. 8, 296–302. New York: Macmillan.

Ellis, G. 2006. On the nature of emergent reality. In *The Re-Emergence of Emergence: The Emergentist Hypothesis from Science to Religion*, ed. P. Clayton and P. Davies, 79–109. Oxford: Oxford University Press.

Farlow, J. O., C. V. Thompson, and D. E. Rosner. 1976. Plates of the dinosaur Stegosaurus: forced convection heat loss fins? *Science* 192: 1123–5.

Findlay, J. N. 1948. Can God's existence be disproved? *Mind* 57: 176–83.

Fischer, J. M., R. Kane, D. Pereboom, and M. Vargas. 2007. *Four Views on Free Will.* Malden, Mass.: Blackwell.

Fodor, J. 1996. "Peacocking." *London Review of Books*, no. 18 (April), pp. 19–20.

Freeman, S., and J. C. Herron. 2004. *Evolutionary Analysis* (third edition). Englewood Cliffs, N.J.: Prentice-Hall.

Freud, S. 1900. *The Interpretation of Dreams.* Translator A. A. Brill. London: Allen and Unwin.

1920. *Dream Psychology: Psychoanalysis for Beginners.* Translator M. D. Eder. New York: James A. McCann Company.

Galileo. 1953. *Dialogue Concerning the Two Chief World Systems.* Translator S. Drake. Berkeley: University of California Press.

Garber, D. 1992. *Descartes' Metaphysical Physics.* Chicago: University of Chicago Press.

Gödel, K. 1995. *Collected Works.* Editors S. Feferman, J. W. Dawson Jr., W. Goldfarb, C. Parsons, and R. M. Solovay. Vol. 3. New York: Oxford University Press.

Goodfield, G. J. 1960. *The Growth of Scientific Physiology: Physiological Method and the Mechanist-Vitalist Controversy, Illustrated by the Problems of Respiration and Animal Heat.* London: Hutchinson.

Goodwin, B. 2001. *How the Leopard Changed Its Spots* (second edition). Princeton, N.J.: Princeton University Press.

Gould, S. J. 1989. *Wonderful Life: The Burgess Shale and the Nature of History*. New York: Norton.

1999. *Rocks of Ages: Science and Religion in the Fullness of Life*. New York: Ballantine.

Gould, S. J., and R. C. Lewontin. 1979. The spandrels of San Marco and the Panglossian paradigm: a critique of the adaptationist programme. *Proceedings of the Royal Society of London, Series B: Biological Sciences* 205: 581–98.

Grant, E. 1996. *The Foundations of Modern Science in the Middle Ages: Their Religious, Institutional and Intellectual Contexts*. Cambridge: Cambridge University Press.

Gray, A. 1876. *Darwiniana*. New York: D. Appleton.

1881. *Structural Botany* (sixth edition). London: Macmillan.

Grünbaum, A. 2007. Why is there a universe at all, rather than just nothing? In *The Routledge Companion to Philosophy of Religion*, ed. C. Copan and P. Meister, 441–51. London: Routledge.

Haeckel E. 1866. *Generelle Morphologie der Organismen*. Berlin: Georg Reimer.

Haldane, J., and P. Lee. 2003. Aquinas on human ensoulment, abortion and the value of life. *Philosophy* 78: 255–78.

Haldane, J. B. S. 1927. *Possible Worlds and Other Essays*. London: Chatto and Windus.

Hale, B., and C. Wright, editors. 2001. *The Reason's Proper Study: Essays Towards a Neo-Fregean Philosophy of Mathematics*. Oxford: Clarendon Press.

Hall, A. R. 1954. *The Scientific Revolution 1500–1800: The Formation of the Modern Scientific Attitude*. London: Longman, Green.

1983. *The Revolution in Science, 1500–1750*. London: Longman.

Harré, R. 1972. *The Philosophies of Science: An Introductory Survey*. Oxford: Oxford University Press.

Haught, J. A. 1996. *2000 years of Disbelief: Famous People with the Courage to Doubt*. Amherst, N. Y.: Prometheus Books.

Hauser, M. D. 2006. *Moral Minds: How Nature Shaped Our Universal Sense of Right and Wrong*. New York: Ecco.

Heath, T. 1963. *A Manual of Greek Mathematics*. Mineola, N.Y.: Dover.

Hegel, G. W. F. [1817] 1970. *Philosophy of Nature*. Trans. A. V. Miller. Oxford: Oxford University Press.

Heidegger, M. 1959. *An Introduction to Metaphysics*. New Haven, Conn.: Yale University Press.

Herschel, J. F. W. 1830. *Preliminary Discourse on the Study of Natural Philosophy*. London: Longman, Rees, Orme, Brown, Green, and Longman.

Hick, J. 1961. Necessary being. *Scottish Journal of Theology* 14: 353–69.

1976. *Death and Eternal Life*. New York: Harper and Row.

1978. *Evil and the God of Love*. New York: Harper and Row.

Hobbes, T. [1651] 1998. *Leviathan*. Edited by J. C. A. Gaslein. Oxford: Oxford University Press.

Holwerda, D. 1983. Faith, reason, and the resurrection in the theology of Wolfhart Pannenberg. In *Faith and Rationality: Reason and Belief in God*, ed. A. Plantinga and N. Wolterstorff, 265–316. Notre Dame: University of Notre Dame Press.

Hume, D. [1779] 1947. *Dialogues Concerning Natural Religion*. Editor N. K. Smith. Indianapolis: Bobbs-Merrill.

[1777] 1975. *Enquiries Concerning the Human Understanding, and Concerning the Principles of Morals*. Editor T. L. Beauchamp. Oxford: Oxford University Press.

[1739] 2000. *A Treatise of Human Nature*. Editors D. F. and M. J. Norton. Oxford: Oxford University Press.

Huxley, J. S. 1942. *Evolution: The Modern Synthesis*. London: Allen and Unwin.

Huxley, T. H. 1857–59. On the theory of the vertebrate skull. Croonian Lecture delivered before the Royal Society, June 17, 1858. *Proceedings of the Royal Society*, 381–457.

1873. *Critiques and Addresses*. New York: Appleton.

1874. On the hypothesis that animals are automata, and its history. In *Collected Essays, Volume I, Methods and Results*, 195–250. London: Macmillan.

James, W. 1880. *The Principles of Psychology*. New York: Henry Holt.

John Paul II. 1997. The Pope's message on evolution. *Quarterly Review of Biology* 72: 377–83.

1998. *Fides et Ratio: Encyclical Letter of John Paul II to the Catholic Bishops of the World*. Vatican City: L'Osservatore Romano.

Johnson-Laird, P. 1988. *The Computer and the Mind*. Cambridge, Mass.: Harvard University Press.

Johnson, M., editor. 1981. *Philosophical Perspectives on Metaphor*. Minneapolis: University of Minnesota Press.

Kane, R. 2007. Libertarianism. In *Four Views about Free Will*, ed. J. M. Fischer, R. Kane, D. Pereboom, and M. Vargas, 5–43. Malden, Mass.: Blackwell.

Kanigel, R. 1991. *The Man Who Knew Infinity: A Life of the Genius Ramanujan*. New York: Charles Scribner's Sons.

Kant, I. [1781] 1929. *Critique of Pure Reason*. Translator N. Kemp Smith. New York: Humanities Press.

[1788] 1948. *Critique of Practical Reason*. Translator T. K. Abbott. London: Longmans, Green.

[1790] 1951. *Critique of Judgement*. Translator J. C. Meredith. New York: Hafner.

[1785] 1959. *Foundations of the Metaphysics of Morals*. Translator C. W. Beck. Indianapolis: Bobbs-Merrill.

Kauffman, S. A. 1995. *At Home in the Universe: The Search for the Laws of Self-Organization and Complexity*. New York: Oxford University Press.

2008. *Reinventing the Sacred: A New View of Science, Reason, and Religion.* New York: Basic Books.

Kellogg, V. L. 1907. *Darwinism Today.* New York: Henry Holt.

Kirchner, J. W. 2002. The Gaia hypothesis: fact, theory, and wishful thinking. *Climatic Change* 52: 391–408.

Kitcher, P. 2007. *Living with Darwin: Evolution, Design, and the Future of Faith.* New York: Oxford University Press.

Kuhn, T. 1957. *The Copernican Revolution.* Cambridge, Mass.: Harvard University Press.

1962. *The Structure of Scientific Revolutions.* Chicago: University of Chicago Press.

1977. *The Essential Tension: Selected Studies in Scientific Tradition and Change.* Chicago: University of Chicago Press.

1993. Metaphor in science. In *Metaphor and Thought* (second edition), ed. Andrew Ortony, 533–42. Cambridge: Cambridge University Press.

La Mettrie, J. O. de. 1912. *Man a Machine.* Chicago: Open Court.

Lakoff, G., and M. Johnson. 1980. *Metaphors We Live By.* Chicago: University of Chicago Press.

Lamarck, J. B. 1802. *Discours d'ouverture du cours de zoologie.* Paris: Maillard.

Leibniz, G. F. W. [1714a] 1965. *Monadology and Other Philosophical Essays.* New York: Bobbs-Merrill.

[1714b] 1973. *Principles of Nature and of Grace Founded upon Reason. Leibniz: Philosophical Writings.* Editor G. H. R. Parkinson. London: J. M. Dent.

Lenoir, T. 1989. *The Strategy of Life: Teleology and Mechanics in Nineteenth Century German Biology.* Chicago: University of Chicago Press.

Lewes, G. H. 1875. *Problems of Life and Mind, First Series. The Foundation of a Creed.* London: Trübner.

Lewis, D. 2004. Void and object. In *Causation and Counterfactuals*, ed. J. Hall, N. Paul, and L. A. Collins, 277–90. Cambridge, Mass.: MIT Press.

Lewontin, R. C., S. Rose, and L. J. Kamin. 1984. *Not in Our Genes: Biology, Ideology and Human Nature.* New York: Pantheon.

Liebig, J. 1842. *Die organische Chemie in ihrer Anwendung auf Physiologie un Pathologie.* Braunschweig: Vieweg und Sohn.

Lindberg, D. C. 1992. *The Beginnings of Western Science: The European Scientific Tradition in Philosophical, Religious, and Institutional Context, Prehistory to A.D. 1450.* Chicago: University of Chicago Press.

1995. Medieval science and its religious context. *Osiris* 10: 61–79.

Lloyd Morgan, C. 1894. *An Introduction to Comparative Psychology.* London: W. Scott.

Locke, J. 1689. *An Essay Concerning Human Understanding.* Editor A. C. Fraser. New York: Dover.

Lovelock, J. E. 1979. *Gaia: A New Look at Life on Earth.* Oxford: Oxford University Press.

2007. What is Gaia? In *Philosophy of Biology* (second edition), ed. M. Ruse, 307–8. Buffalo, N.Y.: Prometheus.

Lowe, E. J. 1997. There are no easy problems of consciousness. In *Explaining Consciousness – The 'Hard Problem'*, ed. J. Shear, 117–24. Cambridge, Mass.: MIT Press.

Lucas, J. R. 1961. Minds, machines and Gödel. *Philosophy* 36: 112–27.

Mackie, J. 1977. *Ethics*. Harmondsworth, Mddx.: Penguin.

Maienschien, J. 2003. *Whose View of Life? Embryos, Cloning, and Stem Cells*. Cambridge, Mass.: Harvard University Press.

Margulis, L., and D. Sagan. 1997. *Slanted Truths: Essays on Gaia, Symbiosis and Evolution*. Seacaucus, N.J.: Copernicus Books.

Mayr, E. 1963. *Animal Species and Evolution*. Cambridge, Mass.: Harvard University Press.

1988. *Towards a New Philosophy of Biology: Observations of an Evolutionist*. Cambridge, Mass.: Belknap.

McGinn, C. 1997. Consciousness and space. In *Explaining Consciousness – The 'Hard Problem'*, ed. J. Shear, 97–108. Cambridge, Mass.: MIT Press.

2000. *The Mysterious Flame: Conscious Minds in a Material World*. New York: Basic Books.

Mehren, E. 1989. The cosmic lottery. *Los Angeles Times*. November 8, E1, E6.

Mendelson, M. 1995. The dangling thread: Augustine's three hypotheses of the soul's origin in "De Genesi ad Litteram." *The British Journal for the History of Philosophy* 3: 219–47.

Merchant, C. 1980. *The Death of Nature: Women, Ecology, and the Scientific Revolution. A Feminist Reappraisal of the Scientific Revolution*. Scranton, Pa.: HarperCollins.

1995. *Earthcare: Women and the Environment*. London: Routledge.

2003. *Reinventing Eden: The Fate of Nature in Western Culture*. London: Routledge.

Mill, J. S. 1840. Review of the works of Samuel Taylor Coleridge. *London Westminster Review* 33: 257–302.

1843. *A System of Logic Ratiocinative and Inductive*. Edited by J. M. Robson. Toronto: University of Toronto Press.

Moravia, S. 1978. From *homme machine* to *homme sensible*: Changing eighteenth-century models of man's image. *Journal of the History of Ideas* 39: 45–60.

Mullin, R. B. 2000. Miracle. In *The Oxford Companion to Christian Thought*, ed. A. Hastings, A. Mason, and H. Pyper, 438–40. Oxford: Oxford University Press.

Murphy, N. 2006. Emergence and mental causation. In *The Re-Emergence of Emergence: The Emergentist Hypothesis from Science to Religion*, ed. P. Clayton and P. Davies, 227–43. Oxford: Oxford University Press.

Murphy, N., and W. S. Brown. 2007. *Did My Neurons Make Me Do It? Philosophical and Neurobiological Perspectives on Moral Responsibility and Free Will*. Oxford: Oxford University Press.

Nagel, E. 1961. *The Structure of Science: Problems in the Logic of Scientific Explanation*. New York: Harcourt, Brace and World.

Newman, J. H. 1973. *The Letters and Diaries of John Henry Newman*. Editors C. S. Dessain and T. Gornall. Vol. 25. Oxford: Clarendon Press.

Niebuhr, R. 1941. *The Nature and Destiny of Man. I. Human Nature*. New York: Charles Scribner's Sons.

Niklas, K. J. 1988. The role of phyllotactic pattern as a 'developmental constraint' on the interception of light by leaf surfaces. *Evolution* 42: 1–16.

O'Connor, T. 2000. *Persons and Causes: The Metaphysics of Free Will*. New York: Oxford University Press.

Owen, R. 1849. *On the Nature of Limbs*. London: Voorst.

Pannenberg, W. 1968. *Jesus – God and Man*. London: SCM Press.

Parfit, D. 1998a. Why anything? Why this? Part 1. *London Review of Books* 20, no. 2 (January 22): 24–7.

1998b. Why anything? Why this? Part 2. *London Review of Books* 20, no. 3 (February 5): 22–5.

Pasnau, R. 2001. *Thomas Aquinas on Human Nature*. Cambridge: Cambridge University Press.

Paul VI. 1968. *Humanae Vitae*. Vatican City: L'Osservatore Romano.

Pavlov, I. P. 1927. *Conditioned Reflexes: An Investigation of the Physiological Activity of the Cerebral Cortex*. Translator G. V. Anrep. Oxford: Oxford University Press.

Pereboom, D. 2007. Hard incompatibilism. In *Four Views on Free Will*, ed. J. M. Fischer, R. Kane, D. Pereboom, and M. Vargas, 85–125. Malden, Mass.: Blackwell.

Pike, N., editor. 1964. *God and Evil*. Englewood Cliffs, N.J.: Prentice-Hall.

1970. *God and Timelessness*. New York: Schocken Books.

Pinker, S. 1997. *How the Mind Works*. New York: Norton.

Plantinga, A. 1980. *Does God Have a Nature?* Milwaukee, Wisc.: Marquette University Press.

1983. Reason and belief in God. In *Faith and Rationality: Reason and Belief in God*, ed. A. Plantinga, and N. Woltersdorff, 16–93. Notre Dame, Ind.: University of Notre Dame Press.

1999. Augustinian Christian philosophy. In *The Augustinian Tradition*, ed. G. B. Matthews, 1–26. Berkeley: University of California Press.

Polkinghorne, J. C. 1986. *One World: The Interaction of Science and Theology*. Princeton, N. J.: Princeton University Press.

1994. *Science and Christian Belief: Theological Reflections of a Bottom-Up Thinker*. London: SPCK.

Provine, W. B. 1987. Review of *Trial and Error: The American Controversy of Creation and Evolution*, by Edward J. Larson. *Academe* 73: 51–2.

Quine, W. V. O. 1981. *Success and the Limits of Mathematization: Theories and Things*. Cambridge, Mass.: Harvard University Press.

Quinn, P. L. 1978. *Divine Commands and Moral Requirements*. Oxford: Clarendon Press.

Rawls, J. 1971. *A Theory of Justice.* Cambridge, Mass. Harvard University Press.

Reeve, H. K., and P. W. Sherman. 1993. Adaptation and the goals of evolutionary research. *Quarterly Review of Biology* 68: 1–32.

Reznick, D. N., M. V. Butler IV, A. Rodd, and P. Ross. Life-history evolution in guppies (Poecilia reticulata) 6. Differential mortality as a mechanism for natural selection. *Evolution* 50: 1651–60.

Reznick, D. N., and J. Travis. 1996. The empirical study of adaptation in natural populations. In *Adaptation*, ed. M. R. Rose and G. V. Lauder. San Diego: Academic Press.

Richards, R. J. 2003. *The Romantic Conception of Life: Science and Philosophy in the Age of Goethe.* Chicago: University of Chicago Press.

2008. *The Tragic Sense of Life: Ernst Haeckel and the Struggle over Evolutionary Thought.* Chicago: University of Chicago Press.

Roberts, A., and J. Donaldson, editors. 1989. *The Ante-Nicene Fathers. Vol. 3.* Grand Rapids, Mich.: Eerdmans.

Roger, J. 1997. *Buffon: A Life in Natural History.* Translator S. L. Bonnefoi. Ithaca, N.Y.: Cornell University Press.

Rundle, B. 2004. *Why There Is Something rather than Nothing.* Oxford: Oxford University Press.

Ruse, M. 1973. *The Philosophy of Biology.* London: Hutchinson.

1975. Darwin's debt to philosophy: an examination of the influence of the philosophical ideas of John F. W. Herschel and William Whewell on the development of Charles Darwin's theory of evolution. *Studies in History and Philosophy of Science* 6: 159–81.

1981. What kind of revolution occurred in geology? *PSA 1978*, 2: 240–73.

1986. *Taking Darwin Seriously: A Naturalistic Approach to Philosophy.* Oxford: Blackwell.

editor. 1988a. *But Is It Science? The Philosophical Question in the Creation/Evolution Controversy.* Buffalo, N.Y.: Prometheus.

1988b. *Homosexuality: A Philosophical Inquiry.* Oxford: Blackwell.

1996. *Monad to Man: The Concept of Progress in Evolutionary Biology.* Cambridge, Mass.: Harvard University Press.

1999a. *The Darwinian Revolution: Science Red in Tooth and Claw* (second edition). Chicago: University of Chicago Press.

1999b. *Mystery of Mysteries: Is Evolution a Social Construction?* Cambridge, Mass.: Harvard University Press.

2001. *Can a Darwinian Be a Christian? The Relationship between Science and Religion.* Cambridge: Cambridge University Press.

2003. *Darwin and Design: Does Evolution Have a Purpose?* Cambridge, Mass.: Harvard University Press.

2005a. *The Evolution-Creation Struggle.* Cambridge, Mass.: Harvard University Press.

2005b. Darwin and mechanism: Metaphor in science. *Studies in History and Philosophy of Biology and Biomedical Sciences* 36: 285–302.

2006. Kant and evolution. In *Theories of Generation*, ed. J. Smith, 402–15. Cambridge: University of Cambridge Press.

2008a. *Charles Darwin*. Oxford: Blackwell.

2008b. *Evolution and Religion: A Dialogue*. Lanham, Md.: Rowman and Littlefield.

Russell, E. S. 1916. *Form and Function: A Contribution to the History of Animal Morphology*. London: John Murray.

Russell, R. J. 2008. *Cosmology: From Alpha to Omega, the Creative Mutual Interaction Between Cosmology and Science*. Minneapolis: Fortress Press.

Ryle, G. 1949. *The Concept of Mind*. London: Hutchinson.

Schelling, F. W. J. von. [1803] 1988. *Ideas for a Philosophy of Nature – as Introduction to the Study of this Science*, second edition. Translators E. E. Harris and P. Heath. Cambridge: Cambridge University Press.

[1797] 2004. *First Outline of a System of the Philosophy of Nature*. Translated by K. Peterson. Albany, N.Y.: State University of New York Press.

Searle, J. 1980. Minds, brains and programs. *Behavioral and Brain Sciences* 3: 417–57.

Sedley, D. 2008. *Creationism and Its Critics in Antiquity*. Berkeley: University of California Press.

Shiva, V. 2000. *Stolen Harvest: The Hijacking of the Global Food Supply*. Cambridge, MA: South End Press.

2005. *Earth Democracy: Justice, Sustainability, and Peace*. Cambridge, Mass.: South End Press.

Singer, P. 2005. Ethics and intuitions. *Journal of Ethics* 9: 331–52.

Skinner, B. F. 1938. *The Behavior of Organisms: An Experimental Analysis*. New York: Appleton-Century-Crofts.

1953. *Science and Human Behavior*. New York: Macmillan.

Smart, J. J. C. 1955. The existence of God. In *New Essays in Philosophical Theology*, ed. A. Flew and A. McIntyre. London: SCM Press.

Smuts, J. C. 1926. *Holism and Evolution*. London: Macmillan.

Spinoza, B. [1677] 1985. *Ethics. The Collected Writings of Spinoza*, translator E. Curley, Vol. 1. Princeton, N. J.: Princeton University Press.

1995. *The Letters*. Editors S. Barbone, L. Rice, J. Shirley, and S. Adler. Indianapolis: Hackett.

Steiner, M. 1975. Platonism and mathematical knowledge. In his *Mathematical Knowledge*, 109–37. Ithaca, N.Y.: Cornell University Press.

Swinburne, R. 1977. *The Coherence of Theism*. Oxford: Clarendon Press.

2005. *Faith and Reason* (second edition). Oxford: Oxford University Press.

Thagard, P. 2005. *Mind: Introduction to Cognitive Science*. Cambridge, Mass.: MIT Press.

Thorndike, E. 1905. *The Elements of Psychology*. New York: A. G. Seiler.

Tooby, J., L. Cosmides, and H. C. Barrett. 2005. Resolving the debate on innate ideas: learnability constraints and the evolved interpenetration of motivational and conceptual functions. In *The Innate Mind: Structure and*

Content, ed. P. Carruthers, S. Laurence, and S. Stich. New York: Oxford University Press.

Trivers, R. L. 1971. The evolution of reciprocal altruism. *Quarterly Review of Biology* **46**: 35–57.

Van Inwagen, P. 1996. Why is there anything at all? *Proceedings of the Aristotelian Society* **70**: 95–110.

Vickers, B. 2008. Francis Bacon, feminist historiography, and the domination of nature. *Journal of the History of Ideas* **69**: 117–46.

Ward, K. 1996. *God, Chance and Necessity*. Oxford: Oneworld.

Watson, A. J., and J. E. Lovelock. 1983. Biological homeostasis of the global environment: the parable of Daisyworld. *Tellus, Series B: Chemical and Physical Meterology* **35**: 284–9.

Weinberg, S. 1977. *The First Three Minutes: A Modern View of the Origin of the Universe*. New York: Basic Books.

1999. A designer universe. *New York Review of Books* **46**, no. 16: 46–8.

2001. *Facing up: Science and Its Cultural Adversaries* Cambridge, Mass.: Harvard University Press.

Westfall, R. S. 1980. *Never at Rest: A Biography of Isaac Newton*. Cambridge: University of Cambridge Press.

Westman, R. S. 1986. The Copernicans and the churches. In *God and Nature: Historical Essays on the Encounter between Christianity and Science*, ed. D. C. Lindberg, and R. L. Numbers, 76–113. Berkeley: University of Californian Press.

Whewell, W. 1837. *The History of the Inductive Sciences*. London: Parker.

1840. *The Philosophy of the Inductive Sciences*. London: Parker.

Whitehead, A. N. 1929. *Process and Reality: An Essay in Cosmology*. New York: Macmillan.

Williams, B. 1973. The Makropulos case: Reflections on the tedium of immortality. In *Problems of the Self*. Cambridge: Cambridge University Press.

Williams, G. C. 1966. *Adaptation and Natural Selection*. Princeton, N. J.: Princeton University Press.

Williams, R. 2000. Ressurection. In *The Oxford Companion to Christian Thought*, ed. A. Hastings, A. Mason, and H. Pyper, 616–18. Oxford: Oxford University Press.

Wilson, E. B. 1908. Biology: A lecture delivered at Columbia University in the series on Science, Philosophy and Art, November 20, 1907. New York: Columbia University Press.

Wilson, E. O. 1975. *Sociobiology: The New Synthesis*. Cambridge, Mass.: Harvard University Press.

1978. *On Human Nature*. Cambridge, Mass.: Harvard University Press.

1980a. Caste and division of labor in leaf cutter ants (hymenoptera formicidae, Atta). I. The overall pattern in Atta sexdens. *Behavioral Ecology and Sociobiology* **7**: 143–56.

1980b. Caste and division of labor in leaf cutter ants (hymenoptera formicidae, Atta). II. The ergonomic optimization of leaf cutting. *Behavioral Ecology and Sociobiology* 7: 157–65.

Wittgenstein, L. 1923. *Tractatus Logico-Philosophicus.* London: Routledge & Kegan Paul.

1965. A lecture on ethics. *The Philosophical Review* 74: 3–12.

Wright, S. 1932. The roles of mutation, inbreeding, crossbreeding and selection in evolution. *Proceedings of the Sixth International Congress of Genetics* 1: 356–66.

INDEX